THE LIBRARY
ST. MARY'S COLLEGE OF MARYLAND
ST. MARY'S CITY, MARYLAND 20686

D1608283

Dynamic Earth Environments

Dynamic Earth Environments

Remote Sensing Observations from Shuttle–*Mir* Missions

edited by

Kamlesh P. Lulla

National Aeronautics and Space Administration
Lyndon B. Johnson Space Center
Office of Earth Sciences
Houston, Texas USA

and

Lev V. Dessinov

Russian Academy of Sciences
Institute of Geography
Laboratory of Remote Sensing
Moscow, Russia

with

Cynthia A. Evans, Patricia W. Dickerson, Julie A. Robinson

Lockheed Martin Space Operations
Lyndon B. Johnson Space Center
Office of Earth Sciences
Houston, Texas USA

John Wiley & Sons, Inc.

New York • Chichester • Weinheim • Brisbane • Singapore • Toronto

This book is printed on acid-free paper. ∞

Copyright © 2000 by John Wiley & Sons, Inc. All rights reserved.

Published simultaneously in Canada.

No part of this publication may be reproduced, stored in a retrieval system or transmitted in any form or by any means, electronic, mechanical, photocopying, recording, scanning or otherwise, except as permitted under Sections 107 or 108 of the 1976 United States Copyright Act, without either the prior written permission of the Publisher, or authorization through payment of the appropriate per-copy fee to the Copyright Clearance Center, 222 Rosewood Drive, Danvers, MA 01923, (978) 750-8400, fax (978) 750-4744. Requests to the Publisher for permission should be addressed to the Permissions Department, John Wiley & Sons, Inc., 605 Third Avenue, New York, NY 10158-0012, (212) 850-6011, fax (212) 850-6008, E-Mail: PERMREQ@WILEY.COM.

This publication is designed to provide accurate and authoritative information in regard to the subject matter covered. It is sold with the understanding that the publisher is not engaged in rendering professional services. If professional advice or other expert assistance is required, the services of a competent professional person should be sought.

Library of Congress Cataloging-in-Publication Data:
Dynamic earth environments : remote sensing observations from Shuttle-Mir missions /edited by Kamlesh P. Lulla and Lev V. Dessinov.
    p. cm.
  Includes bibliographical references and index.
  ISBN 0-471-39005-4 (cloth : alk. paper)
  1. Remote sensing in earth sciences—International cooperation. 2. Earth—Photographs from space. 3. Endeavour (Space shuttle) 4. Mir (Space station) I. Lulla, Kamlesh. II. Dessinov, Lev V.

QE33.2.R4 D96 2000
551—dc21
                                                                                                                                 00-026818

Printed in the United States of America.

10 9 8 7 6 5 4 3 2 1

Contents

Foreword F. Culbertson and V. V. Ryumin	vii
Preface K. P. Lulla and L. V. Dessinov	ix
Contributors	xiii

CHAPTER 1: Shuttle–*Mir* Earth Science Investigations: Studying Dynamic Earth Environments from the *Mir* Space Station — 1
C. A. Evans, K. P. Lulla, L. V. Dessinov, N. F. Glazovskiy, N. S. Kasimov, and Yu. F. Knizhnikov

CHAPTER 2: Russian Visual Observations of Earth: Historical Perspective — 15
N. F. Glazovskiy and L. V. Dessinov

CHAPTER 3: Twenty-Eight Years of Urban Growth in North America Quantified by Analysis of Photographs from Apollo, Skylab, and Shuttle–*Mir* — 25
J. A. Robinson, B. H. McRay, and K. P. Lulla

CHAPTER 4: Fluctuating Water Levels as Indicators of Global Change: Examples from Around the World — 43
C. A. Evans, J. Caruana, D. L. Amsbury, and K. P. Lulla

CHAPTER 5: The 1997–1998 El Niño: Images of Floods and Drought — 61
C. A. Evans, J. A. Robinson, M. J. Wilkinson, S. Runco, P. W. Dickerson, D. L. Amsbury, and K. P. Lulla

CHAPTER 6: Imaging Aerosols from Low Earth Orbit: Photographic Results from the Shuttle–*Mir* and Shuttle Programs — 77
M. J. Wilkinson, J. D. Wheeler, R. J. Charlson, and K. P. Lulla

CHAPTER 7: Biomass Burning and Smoke Palls with Observations from the Space Shuttle and Shuttle–*Mir* Missions — 99
M. J. Wilkinson, K. P. Lulla, and M. Glasser

CHAPTER 8:	Windows of Opportunity: Photo Survey of the *Mir* Earth Observation Windows *P. B. Saganti and K. P. Lulla*	121

THE CASPIAN SEA — 131

CHAPTER 9:	Geographical, Geological, and Ecological Effects of Caspian Sea-Level Fluctuations: Introduction *N. F. Glazovskiy and V. A. Rudakov*	133
CHAPTER 10:	A Caspian Chronicle: Sea-Level Fluctuations Between 1982 and 1997 *P. W. Dickerson*	145
CHAPTER 11:	Morphological and Geological Structure of the Northern Coast of the Caspian Sea *L. B. Aristarkhova, A. A. Svitoch, and O. N. Bratanova*	149
CHAPTER 12:	Shoreline Dynamics and the Hydrographic System of the Volga Delta *N. I. Alekseevskiy, D. N. Aibulatov, and S. V. Chistov*	159
CHAPTER 13:	Changes in Avian Habitats in Volga Delta Wetlands During Caspian Sea-Level Fluctuations *E. A. Baldina, I. A. Labutina, G. M. Rusanov, A. K. Gorbunov, A. F. Zhivoglyad, and J. de Leeuw*	171
CHAPTER 14:	Changes in Coastal Vegetation in the Northern Caspian Region During Sea-Level Rise *V. I. Kravtsova and E. G. Myalo*	181
CHAPTER 15:	Dynamics of the Northeastern Caspian Sea Coastal Zone in Connection with Sea-Level Rise *V. I. Kravtsova*	191
CHAPTER 16:	Evolution of the Gulf of Kara-Bogaz-Gol in the Past Century *A. N. Varushchenko, S. A. Lukyanova, G. D. Solovieva, A. N. Kosarev, and A. V. Kurayev*	201
CHAPTER 17:	Eddy Formation in the Caspian Sea *L. M. Shipilova*	211
CHAPTER 18:	Geomorphology of Southern Azerbaijan and Coastal Responses to the Caspian Transgression *E. I. Ignatov and G. D. Solovieva*	221
CHAPTER 19:	Land-Use Changes in the Northwest Caspian Coastal Area, 1978–1996: Case Study of the Republic of Kalmykia *A. S. Shestakov*	231
	Appendix: Astronauts and Cosmonauts	241
	Index	253
	Photo insert	261

Foreword

Frank Culbertson
NASA Johnson Space Center, Houston, Texas USA

Valeri V. Ryumin
RSC-Energia, Moscow, Russia

The Phase 1 program began in early 1994 and consisted of a series of Shuttle–*Mir* rendezvous and docking missions and stays of seven astronauts on the Russian *Mir* space station. Nine Russian cosmonauts flew on the Space Shuttle during Phase 1. This program has allowed the United States and its international partners the opportunity for living, working, and conducting long-term research in space. In this context, one of the research activities of choice for the flight crews was to conduct earth observations from both the *Mir* space station and the Space Shuttle windows. The long-term stays on *Mir* made it possible for the crews to observe, photograph, and document Earth surface dynamics and processes over the same site repeatedly. These research opportunities provided a new dimension for photographic data gathering over global sites. The imagery collected during this period has added to the voluminous Earth imagery database already assembled by the Office of Earth Sciences at Johnson Space Center.

We are delighted with the success of the cooperative initiatives in the Earth Sciences during Phase 1 (Shuttle–*Mir*/NASA–*Mir* programs, 1994–1998) of the International Space Station program. We congratulate both the American and Russian teams of scientists, engineers, and managers who accomplished this important scientific activity during this phase. This publication is a valuable record of these joint efforts. We hope that these scientific contributions will enhance our understanding of the changes—both short term and long term—that are occurring on our planet.

In the same spirit of cooperation that was initiated during the first phase, NASA and its international partners are now building the new International Space Station. We expect that this new orbital laboratory will provide robust facilities for future Earth science remote sensing from the windows as well as attached platforms. The International Space Station will play an important and exciting role in Earth science research, just as the earlier U.S. space station Skylab and the Russian space stations Salyut and *Mir* played in the past. The publication of this book marks a beginning of a new era in human-directed Earth remote sensing from space.

Preface

Kamlesh P. Lulla
NASA Johnson Space Center, Houston, Texas USA

Lev V. Dessinov
Russian Academy of Sciences, Moscow, Russia

We are delighted to present this book as solid evidence of the successful completion of the cooperative initiatives in the Earth Sciences during Phase 1 (Shuttle–*Mir* program, 1994–1998) of the International Space Station program.

In 1975, the first Earth Observations from Space cooperative effort between the United States and Russia (then the Soviet Union) was accomplished with the completion of the Apollo–Soyuz Test Project. One of the scientific objectives of the mission was "to gather photographic and observational data in support of ongoing research in the broad fields of geology, hydrology, and oceanography as well as observational data pertaining to meteorology" (*NASA Fact-Sheet 75-9* issued on January 1, 1975). This historic milestone was a guiding beacon that led to the joint U.S.–Russian efforts to conduct Earth science during the Shuttle–*Mir* program.

A new era of international cooperation in human spaceflight between the United States and Russia began in February 1994 with the launch of Space Shuttle *Discovery* (STS-60). The astronaut crew included five American astronauts—Charlie Bolden, Ron Sega, Ken Reightler, Jan Davis, and Franklin Chang-Diaz—and the first Russian cosmonaut—Sergei Krikalev. This flight also marked the beginning of joint efforts to conduct Earth science and observations. One of the highlights of this flight was a film test to assess the suitability of U.S. and Russian infrared films for Earth photography during the long-duration missions. This initial effort was expanded to include Earth science as an experiment on all subsequent Shuttle–*Mir* missions. The broad objectives of this joint effort in Earth science remained the same as those of the 1975 Apollo–Soyuz Test Project.

The specific objectives of the Shuttle–*Mir* Earth science efforts were to document change on Earth's surface, especially environmental changes and dynamic processes. These image data have been assimilated into the larger database of Earth photographs taken by astronauts and cosmonauts and will be used for new and continuing studies on global changes.

The database of images is an international public-domain resource for the global science community. The Shuttle–*Mir* imagery, alone and as a part of the larger photographic database from other missions, is a valuable contribution toward understanding the dynamic Earth environments.

We are pleased to note that this publication describes the results of this joint endeavor. The authors whose papers appear in this volume are to be commended for their extraordinary efforts. They have applied the imagery acquired from these joint missions to their studies and analyses and have written substantive manuscripts that make this volume extremely valuable.

We are especially pleased to note that papers by our Russian colleagues have focused on the Caspian region. This regional focus complements the global observations discussed in the first section. These chapters provide excellent examples of the importance and value of space-based Earth observations in regional applications at a variety of scales.

We deeply acknowledge excellent support from both internal and external reviewers. The peer-review process was made smooth by full cooperation from our external reviewers. We thank the team of reviewers from Russia for their valuable inputs in this process.

We especially thank Dr. Michael Helfert, Director, Southeast Regional Climate Center, Columbia, South Carolina for reviewing a large set of manuscripts and for providing valuable insights into the Russian Earth science activities. We also thank Professor Vic Klemas of the University of Delaware, Dr. David Pitts of the University of Houston Clear Lake, and Professor Roger White of the Memorial University of Newfoundland for reviewing selected papers.

Our gratitude to all the members of the flight crews—both U.S. and Russian—who were in orbit during the 11 Space Shuttle flights and seven *Mir* missions that constituted the Phase 1 program. They acquired the images that form the basis of this book. Their scientific acumen and technical expertise resulted in a voluminous collection of Earth images. Also, the acquisition of these data would not have been possible without the hard work and professional expertise supplied by the entire Phase 1 operations and integration team. We thank Information Services Office at Johnson Space Center for support in processing and archiving the images.

We also acknowledge continued support from the members of NASA/Johnson Space Center management. Especially, the support of Mr. Frank Culbertson, manager of the Phase 1 Program Office, and Dr. John Uri, Phase 1 Mission Scientist, is gratefully acknowledged. We also appreciate the efforts of Dr. Peggy Whitson, Dr. John Charles, and Dr. Tom Sullivan in supporting our activities.

We thank Dr. Doug Blanchard, Chief, Earth Science and Solar System Exploration Division; Dr. David Williams, M.D., Director, Dr. John Rummel, Associate Director, and Mr. Richard Nygren, Assistant Director, all of the Space and Life Science Directorate for their encouragement and support throughout these Phase 1 efforts.

It has been our pleasure to work with our colleagues Dr. Cynthia A. Evans, Dr. Patricia Wood Dickerson, and Dr. Julie Robinson of the Office of Earth Sciences, who served as the Associate Editors of this volume. We appreciate their dedication and hard work on this publication. Dr. Justin Wilkinson of the Office of Earth Sciences provided crucial support during the editing process. The entire staff of the Office of Earth Sciences provided assistance during assembly of the book. Our thanks to all of them. We especially thank Sue Runco, Joe Caruana, Kim Willis, Leslie Upchurch, and Marco Lozano for their contributions.

It is our hope that this publication will excite you to explore our own home planet!

Contributors

D. N. AIBULATOV
Department of Geography, Moscow State University Moscow, 119899, Russia

N. I. ALEKSEEVSKIY
Department of Geography, Moscow State University Moscow, 119899, Russia

DAVID L. AMSBURY
Office of Earth Sciences, NASA Lyndon B. Johnson Space Center, Houston, TX 77058, USA.

L. B. ARISTARKHOVA
Department of Geography, Moscow State University Moscow, 119899, Russia

E. A. BALDINA
Department of Geography, Moscow State University, 119899, Moscow, Russia

O. N. BRATANOVA
Department of Geography, Moscow State University Moscow, 119899, Russia

JOE CARUANA
Lockheed Martin Space Operations and Office of Earth Sciences, NASA Lyndon B. Johnson Space Center, Houston, Texas 77058, USA.

R. J. CHARLSON
Department of Atmospheric Sciences, University of Washington, Box 351640, Seattle, WA 98195-1640, USA.

S. V. CHISTOV
Department of Geography, Moscow State University Moscow, 119899, Russia

J. DE LEEUW
International Institute for Aerospace Survey and Earth Sciences (ITC), Agriculture, Conservation and Environment Division, P.O. Box 6, 7500 AA Enschede, The Netherlands

LEV V. DESSINOV
Institute of Geography, Russian Academy of Sciences, Moscow 109017, Russia

PATRICIA W. DICKERSON
Lockheed Martin Space Operations and Office of Earth Sciences, NASA Lyndon B. Johnson Space Center, Houston, Texas 77058, USA.

CYNTHIA A. EVANS
Lockheed Martin Space Operations and Office of Earth Sciences, NASA Lyndon B. Johnson Space Center, Houston, Texas 77058, USA.

MARVIN GLASSER
Department of Physics & Physical Sciences, University of Nebraska, Kearney, Nebraska 68849, USA.

NIKITA F. GLAZOVSKIY
Institute of Geography, Russian Academy of Sciences, Moscow 109017, Russia

A. K. GORBUNOV
Astrakhanskiy Biosphere Reserve, Naberezhnaya r. Tzarev, 119 Astrakhan 414021, Russia

E. I. IGNATOV
Department of Geography, Moscow State University Moscow, 119899, Russia

N. S. KASIMOV
Institute of Geography, Russian Academy of Sciences, Moscow 109017, Russia

YU F. KNIZHNIKOV
Institute of Geography, Russian Academy of Sciences, Moscow 109017, Russia

A. N. KOSAREV
Department of Geography, Moscow State University, Moscow, 119899, Russia

VALENTINA I. KRAVTSOVA
Department of Geography, Moscow State University, Moscow, 119899, Russia

A. V. KURAEV
Department of Geography, Moscow State University, Moscow, 119899, Russia

I. A. LABUTINA
Department of Geography, Moscow State University, 119899, Moscow, Russia

S. A. LUKYANOVA
Department of Geography, Moscow State University, Moscow, 119899, Russia

KAMLESH P. LULLA
Office of Earth Sciences, Mail Code SN13, NASA Lyndon B. Johnson Space Center, Houston, TX 77058, USA.

BRETT MCRAY
Lockheed Martin Space Operations and Office of Earth Sciences, NASA Lyndon B. Johnson Space Center, Houston, Texas 77058, USA.

ELENA G. MYALO
Department of Geography, Moscow State University, Moscow, 119899, Russia

JULIE A. ROBINSON
Lockheed Martin Space Operations and Office of Earth Sciences, NASA Lyndon B. Johnson Space Center, Houston, Texas 77058, USA.

V. A. RUDAKOV
Russian Academy of Sciences and Department of Geography, Moscow State University, 119899, Moscow, Russia

SUSAN K. RUNCO
Office of Earth Sciences, NASA Lyndon B. Johnson Space Center, Houston, Texas 77058, USA.

G. M. RUSANOV
Astrakhanskiy Biosphere Reserve, Naberezhnaya r. Tzarev, 119 Astrakhan 414021, Russia

PREMKUMAR B. SAGANTI
Image Science and Analysis Group, Lockheed Martin Space Operations, Mail Code C23, 2400 NASA Rd. 1, Houston TX 77058, U.S.A.

A. S. SHESTAKOV
Institute of Geography, Russian Academy of Sciences, Moscow 109017, Russia

L. M. SHIPILOVA
Department of Geography, Moscow State University, Moscow, 119899, Russia

G. D. SOLOVIEVA
Department of Geography, Moscow State University Moscow, 119899, Russia

A. A. SVITOCH
Department of Geography, Moscow State University Moscow, 119899, Russia

A. N. VARUSHCHENKO
Department of Geography, Moscow State University, Moscow, 119899, Russia

M. J. WILKINSON
Lockheed Martin Space Operations and Office of Earth Sciences, NASA Lyndon B. Johnson Space Center, Houston TX 77058, USA.

J. D. WHEELER
Department of Atmospheric Sciences, University of Washington, Box 351640, Seattle, WA 98195-1640, USA.

A. F. ZHIVOGLYAD
Astrakhanskiy Biosphere Reserve, Naberezhnaya r. Tzarev, 119 Astrakhan 414021, Russia

FOREWARD CONTRIBUTORS

FRANK CULBERTSON
NASA Astronaut and Shuttle–*Mir* Program Manager. NASA Lyndon B. Johnson Space Center, Houston, TX 77058 USA

VALERIE V. RYUMIN
Russian Cosmonaut and Director of the Russian portion of the Shuttle–*Mir* program, RSC-Energia, Moscow, Russia

Dynamic Earth Environments

Chapter

1

Shuttle–*Mir* Earth Science Investigations: Studying Dynamic Earth Environments from the *Mir* Space Station

Cynthia A. Evans and Kamlesh P. Lulla

Office of Earth Sciences
NASA Johnson Space Center
Houston, Texas USA

Lev V. Dessinov, N. F. Glazovskiy, N. S. Kasimov, and Yu. F. Knizhnikov

Institute of Geography
Russian Academy of Sciences
Moscow, Russia

ABSTRACT

The joint United States–Russian experiment, Visual Observations of Earth, performed on the Mir Space Station between March 1996 and June 1998 returned more than 22,000 photographs of Earth. These photographs document long-term study sites and dynamic events on the Earth's surface. The overall characteristics of the imagery are described.

1.1 INTRODUCTION

The Johnson Space Center's Office of Earth Sciences and the Institute of Geography of the Russian Academy of Sciences participated jointly in the Shuttle–*Mir* science program aboard the Russian *Mir* space station from March 1996 through June 1998. The prime earth science experiment was Visual Observations of the Earth. *Mir*-based astronauts and cosmonauts photographed Earth's surface throughout their long-duration missions.

The primary goal of Visual Observations of the Earth was to use astronaut and cosmonaut photographs to document environmental changes and dynamic Earth processes such as flooding and droughts, urban growth and land-use changes around the world, events related to El Niño, and transient phenomena such as tropical storms, large fires, and volcanic eruptions. The main difference between the Shuttle- and *Mir*-based programs is that crews of long-duration missions observe and record a *continuum* of changes on Earth, including seasonal changes. The Shuttle–*Mir* photographs are assimilated into the larger database of Earth photographs taken by astronauts and cosmonauts, and are used for new and continuing studies of global change (e.g., Lulla et al., 1996).

A second important objective of the Shuttle–*Mir* visual observations experiment was to use an operational environment to develop approaches and tools for the next generation of Earth observations from the International Space Station (ISS). The Shuttle–*Mir* missions have served well to prepare the NASA Office of Earth Sciences for long-duration scientific investigations from the ISS. Training and operational support developed and tested on the *Mir* missions include interactive electronic training and reference software and a JSC-based interactive map and planning program for ingesting *Mir* vector and lighting conditions to allow planning for site photography. Together with the crews, we have also learned important changes in operations, with implications about the training and equipment provided to the crew while on orbit. The *Mir* missions have allowed us to evaluate the long-term impact of such relatively new on-orbit procedures as film reloading, film management, data recording, and camera maintenance (Figure 1.1). Perhaps most important, we have learned to plan and communicate effectively with the crew from remote centers.

1.2 HISTORY OF ASTRONAUT OBSERVATIONS OF EARTH

Earth science from long-duration space missions is not new to either the United States or Russia. Both countries have solid experience and historical archives for

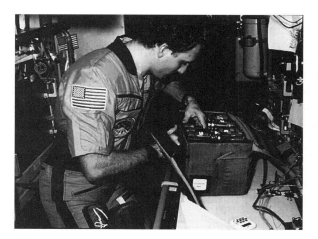

Figure 1.1 Jerry Linenger in the *Mir* Spektr module packing 70-mm film (NM23-59-028). Film and data management are especially important on long-duration missions.

documenting changes on Earth's surface from low Earth orbit. Both programs maintained extensively coordinated and instrumented programs in the 1970s (NASA's Skylab, early Soviet Salyuts, and the joint Apollo–Soyuz test program).

Soviet Earth observations began in the 1960s with pilot experiments on early Soyuz spacecraft. The programs gradually ramped up through the Salyut missions in the 1970s. By Salyut 5 in 1976, the Soviets clearly had demonstrated the potential applications of Earth imagery from long-duration space missions to environmental studies (Koval and Desinov, 1987) (see Chapter 9). With this experience, large-scale programs were launched on Salyut 6 and 7, and for almost a decade (1977–1986), the Soviet program, using both large format and multispectral stationary cameras and hand-held photography by cosmonauts, returned tens of thousands of images each mission. These data were incorporated into the research of hundreds of scientists in the former Soviet Union and other countries (Koval and Desinov, 1987) (Chapter 9). Another generation of observations was established during the early *Mir* years (Chapter 9).

The NASA Earth observations program also began in the 1960s, including the Gemini Earth Photography experiments (Ewing, 1965; Foster and Smistad, 1967; Lowman and Teidemann, 1971; Amsbury, 1989), experiments on Apollo and Apollo–Soyuz (El-Baz and Warner, 1979; Lowman, 1985), and an extensive program on the Skylab station, which included both hand-held and mounted multispectral photography (Anonymous, 1977). Long-duration observations of Earth by U.S. astronauts ended with Skylab, but frequent observations of Earth using hand-held cameras have continued from 1981 to the present on the Space Shuttle flights (Helfert and Lulla, 1989; Lulla et al., 1996).

1.3 METHODS

The scientific approach for *Mir*-based visual observations experiment was simple. NASA provided to the crews Hasselblad 70-mm cameras (equipped with 50-, 100-, and 250-mm lenses), a Nikon F3 35-mm camera, and ample film. Jointly, Russian and U.S. scientists created a list of sites on Earth's surface and requested photography of the sites when conditions allowed. Prior to each flight, scientists from various earth science disciplines trained the crew on important issues and recognition of features and processes. A key component of the approach was crew initiative. Through our collective experiences from the Space Shuttle and earlier *Mir* and Salyut missions, we know that some of the best views and most interesting phenomena cannot be anticipated but can be documented by well-trained astronaut observers (Figure 1.2). Events occur on different time scales, and crews on *Mir* observe and record Earth dynamics daily, weekly, monthly, and seasonally.

Scientists on the ground communicated with the Shuttle–*Mir* crews each week. Those weekly updates comprised a synopsis of significant current events, such as floods, storms, or burning and a prioritized list of sites and their overpass times for each day. The list included planned sites as well as targets of opportunity. The updates were intended to be used as a guide for the crew's photography of Earth.

1.4 RESULTS

Although modest compared to experiments in the 1970s, the Shuttle–*Mir* missions have returned more than 22,000 images of Earth, taken between March 1996 and June

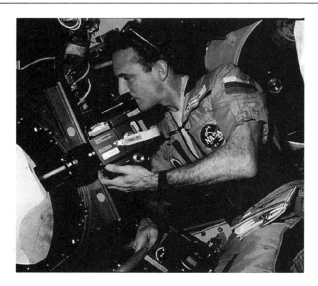

Figure 1.2 Astronaut Jerry Linenger in Kvant-2 photographing Earth with the Hasselblad 70-mm camera (NM23-55-008).

1998 (Figure 1.3). There are several examples of images in the *Mir* data sets (cataloged as missions NM21, NM22, NM23, NASA 5, NASA 6, and NASA 7) that are unique to the NASA JSC Office of Earth Sciences database, which includes more than 300,000 images of Earth taken by NASA astronauts. The Shuttle–*Mir* data indicate that long-duration missions provide several advantages for human-directed Earth observations over the shorter Shuttle missions. We have observed that individual photographic techniques and sense of geography improve rapidly on orbit. This includes the ability to see and recognize new sites and unique photographic opportunities on Earth's surface.

Land-use change was a major focus of the target sites originally planned for Shuttle–*Mir* observations. These included many environmental situations that are

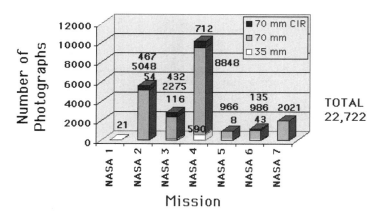

Figure 1.3 Graph showing numbers of images, by film type, acquired during each Shuttle-*Mir* mission.

well studied by the Russian scientists, such as the desiccation of the Aral' Sea, the flooding of the Caspian Sea, and the regional impacts of these events. Several target sites were planned in the Ukrainian and Russian agriculture and industrial belts to document regional impacts of land-use changes resulting from geopolitical events such as the breakup of the Soviet Union, the changing central Asian economy, and continuing recovery at Chernobyl. On other continents, target sites emphasized regions undergoing dramatic changes in land use, such as forest conversion in the Amazon basin, biomass burning and other changes in parts of Africa and South Asia, and expansion of urban regions worldwide.

Seasonal change and long-term climatic effects can readily be documented by looking at water bodies. Regional water issues around the world were deemed of high priority because they are factors in climate-change investigations and directly affect regional land use, local ecology, and the quality of human life. We targeted all phases of water from snow levels and glacier endpoints in high mountain regions to reservoirs, lakes, and inland seas with large fluctuations in water levels. High mountain regions included the Pamir, Karakoram, and Tien Shan mountains (the main source of water for central Asia) and the southern Andes. Important reservoirs included the Aral' and Caspian Seas, Great Salt Lake, Lake Chad, Lake Nasser and the Nile River system, and the central Andean lakes.

We planned to take advantage of the longer-duration missions to document dynamic events over periods equal to or shorter than seasons. A relatively new focus was on aerosol production (dust, smog, smoke) around the world. These data are becoming increasingly important in climate-change modeling, material transport, and land-use change. Regions that we targeted as places to watch for atmospheric events such as dust storms and the production of aerosol blankets include the Sahara and central Andes (dust); the Red Basin of China, the eastern U.S. urban zone, western Europe, and the South Africa industrial core (smog); and the Amazon basin during burning season and the Kalahari (smoke).

We also chose sites with short-term natural dynamics. Ocean processes such as plankton blooms, ocean frontal systems, internal waves, and current boundaries were to be documented wherever observed in the equatorial Pacific, the Arabian Sea, the Gulf Stream, and at convergences of currents off the Argentinean coasts. We also targeted some of the world's most dynamic coastlines, including the Mississippi–Atchafalaya and Yellow River delta systems.

Several of the most active and rapidly changing volcanic regions were chosen for observations as well. These regions included the Kamchatkan Peninsula, Russia (Figure 1.4b); Mt. Pinatubo in the Philippines; the Andes; North Island, New Zealand; the Cascade Mountains of Washington and Oregon; and the central Mexican volcanic belt (Figure 1.4a).

1.4.1 Mission Results

Photography of Earth varied in extent and focus for each of the Shuttle–*Mir* missions. A summary of significant or noteworthy scenes from each mission is provided in Table 1.1. During *Mir* 21/NASA 2 (March to September 1996), astronaut Shannon Lucid and cosmonauts Yuri Usachov and Yuri Onufrienko actively documented the transition from winter to spring to summer in the northern hemisphere. High-

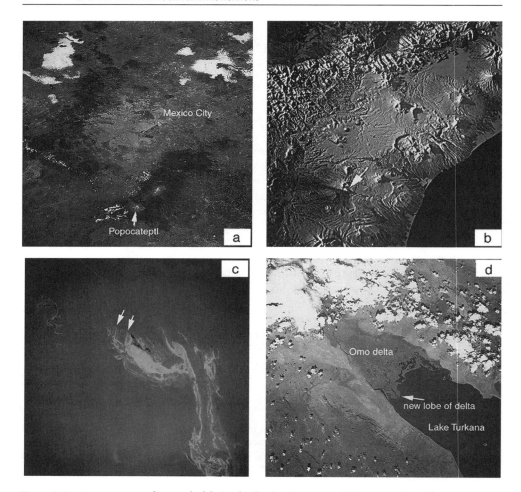

Figure 1.4 Dynamic events photographed during the Shuttle–*Mir* missions. (*a*) Popocatépetl and Mexico City (NASA photograph NM22-727-094). Popocatépetl volcano has been active frequently since 1994. (*b*) Karymsky volcano (NASA photograph NM23-735-992). Kamchatka produces a dark ash layer on top of the surrounding snow cover. (*c*) Oil spill in the Caspian Sea, just offshore from Baku (NASA photograph NM21-773-060A). The arrows mark streamers of oil from individual offshore platforms. The area of the spill, which extends to the north and south from this scene, was greater than 300 km². (*d*) Color-infrared photograph of the Omo River delta building into the north end of Lake Turkana, Kenya (NASA photograph NM22-746-081). Upstream land-use changes and resulting erosion have resulted in a tremendous increase in the size of the delta in the past 20 years. The new lobe along the western edge of the delta did not exist in 1993.

lights included the huge, out-of-control forest fires on the Mongolian steppes in April 1996, and in May and June, less extensive fires along the Russian–Chinese border, and in the Kalahari desert. They also photographed drying reservoirs in the southwestern United States and northern Mexico which indicated the severity of the extensive drought in western North America. NASA 2 photography also included dust storms throughout the spring and summer in the Middle East and over North

Africa, plankton blooms in the North Atlantic Ocean, and continued flooding and a large oil spill in the Caspian Sea (Figure 1.4c).

From September 1996 to January 1997 (NASA 3), astronaut John Blaha and his crewmates Valery Korzun and Alexander Kalery witnessed massive flooding in the lower Nile, continued drought in southern Africa, unusual flooding of dry lake beds in the central Andes, and the spring thaw in the southern Andes. We also point out that the NASA 2 and NASA 3 imagery holds a special significance by way of recording the baseline conditions leading up to the 1997 record-breaking El Niño. For the first time, we have a global picture of pre–El Niño conditions.

From February to May 1997 (NASA 4), Jerry Linenger, Vasili Tsibliyev, and Alexander Lazutkin documented the snow and ice cover over the northern United States and Canada and then tracked lake and sea ice breakup in the Great Lakes, St. Lawrence Seaway, off the Canadian Atlantic coast, and in the Sea of Okhotsk. For three days in February, Linenger photographed and tracked large dust storms over the Tibetan Plateau, a phenomenon rarely documented. In March, they tracked the Ohio–Mississippi River floods. The crew obtained spectacular images of the widespread forest fires in far-eastern Asia (Mongolia, China, and Russia) in late April and early May, and the beginning effects of weather changes related to El Niño in the Andean Altiplano. Other documented events included agricultural burning in Argentina, several dust storms, and winter smog over Europe. Spectacular regional views over Europe, facilitated by the longer mission, which allowed the crew to wait out weather systems, included detailed imagery of western European river systems (Garonne, Loire, Rhône, Rhine, Danube).

The Progress collision in late June 1997 curtailed Earth observations by NASA 5 astronaut Mike Foale and cosmonauts Anatoly Soloviev and Pavel Vinogradov. When able to resume in late July and August 1997, Foale could only photograph the southern hemisphere (lighting constraints and crucial communication passes in the northern hemisphere conflicted with his observations). Despite these on-orbit problems, the crew recorded key atmospheric dynamics related to the developing 1997 El Niño event (thick smoke and haze over South America and Africa), large African dust storms, and new views of aerosol blankets off southern Africa. These data form an important baseline for tracking El Niño impacts on subsequent flights. In this regard, even under adverse conditions, the NASA 5 crew collected an unprecedented data set.

Perhaps the most striking data brought back by Foale from the NASA 5 mission was video imagery of noctilucent clouds. He recorded several minutes of footage of these unusual high-altitude features, and provided commentary. He noted that the crew first noticed them over Siberia, and then the feature appeared to move eastward. Foale's footage begins in late July with coverage over Alaska, but on successive days captures the noctilucent clouds over Canada, and finally over Europe. To our knowledge, this is the first detailed video documentation of these features from low Earth orbit.

David Wolf, the NASA 6 crew member with Soloviev and Vinogradov, continued El Niño observations from September 1997 through January 1998. Highlights included smoke and haze over Sumatra and New Guinea, lake levels in the high Andes, and noteworthy photographs of the unusually lush coast of Somalia after record-breaking rains in northeast Africa. Other imagery acquired during NASA 6 captured dust events in the Middle East and an extensive smog pall over Italy which

TABLE 1.1 Summary of Significant Earth Photographs from Shuttle–Mir Missions

	NASA 2/Mir 21	NASA 3/Mir 22	NASA 4/Mir 22 and 23	NASA 5	NASA 6	NASA 7
Mission dates	March 1 to September 26, 1996	September 19, 1996 to January 22, 1997	January 12 to May 24, 1997	May 15 to October 5, 1997	September 25, 1997 to January 31, 1997	January 22, 1997 to June 12, 1998
Crew	Shannon Lucid, Yuri Usachev, Yuri Onufrienko	John Blaha, Valery Korzun, Alexander Kalery	Jerry Linenger, Vasili Tsibliyev, Alexander Lazutkin	Michael Foale, Vasili Tsibliyev, Alexander Lazutkin	David Wolf, Anatoly Soloviev, Pavel Vinogradov	Andrew Thomas, Talgat Musabayev, Nikolai Budarin
Total number of Earth images	5569	2823	10,150	966	1164	2021
Land-use change	• Saudi Arabian agriculture • Israeli–Egyptian land use • Kuwaiti land use and oil fire recovery • Iraqi drainage in coastal swamps • Lake Chad vegetation • Rondonia, Brazil forest conversion • Yucatan • Russia: Urals mining and industry • Russia: Volga River agriculture • Ukraine: Chernobyl • Black Sea region (Danube, Dniepr, Don) agriculture, industry • Central Asian agriculture (Amu Darya and Syr Darya)	• Land use in Saudi Arabia, Indonesia, Philippines • Kuwaiti land use and oil fire recovery • Yellowstone National Park fire scars • Omo River delta growth, Lake Turkana • Okavango swamp, Botswana • Sudd swamp, Sudan • Lake Victoria • Ganges River system • Indus River agriculture, canals	• Saudi Arabian agriculture • Shatt al Arab • Sudd swamp, Sudan • Nile cultivation • Okavango swamp, Botswana • Amu Darya and Syr Darya agriculture • Indus River agriculture, canals • South American land use: Argentina, Brazil • Western Europe land use • Major urban regions on all continents • Cambodia (Tonle Sap)	• South African land use • Gulf of Fonseca mariculture • Virunga Mountains, Zaire–Rwanda	• Takla Makan agriculture • Israeli–Jordanian agriculture, salt farms • Nile delta agriculture and industry • Aral' Sea agriculture • Lake Chad • Venezuela land use • Libyan agriculture, oil fields, and oases • Urban regions in the Americas • Texas coast • Florida land use	• Takla Makan agriculture • Israeli–Jordanian agriculture, salt farms • Nile delta agriculture and industry • Aral' Sea agriculture • Lake Chad • Urban regions in the Americas • Texas coast • Florida land use • Western plains of the United States • Northern Argentina: agriculture • South-central Russia: land use • Saudi Arabian agriculture • Australian land use

Seasonal change and long-term climate change	• Snowpack and glacier level in Karakoram and Pamirs • Glaciers in Southern Andes • Snowlines across central Asia • Great Lakes ice pack • Hudson Bay icepack • Greenland icepack • Drought in southwestern North America	• Water levels in the Sahel (Inland delta of Niger, Lake Chad, Lake Nasser) • East African lakes • Ice floes off Patagonia • Snow levels in the southern Andes, Alps, Caucasus • General snow blanket over the northern hemisphere (United States, Europe, central Asia, China) • Water levels in the southwestern United States and Brazil • Water levels in the high Andes	• Lake ice, Great Lakes • Sea ice, St. Lawrence Seaway and North Atlantic Ocean • Sea ice, northwest Pacific Ocean • Snow levels, glaciers in the southern Andes Alps, Caucasus • General snow blanket over the northern hemisphere (United States, Europe, Asia) • Flooding of Ohio and Mississippi Rivers • Water levels in Sahel (Inland delta of Niger, Lake Chad, Lake Nasser) • Water levels, southwestern United States, Texas • Water levels, Andes • Water levels, Caspian and Aral′ • Water levels, Tonle Sap, Cambodia	• Lake levels in high Andes (El Niño) • Cloud cover off coastal Peru • Water levels in South Africa • Snowpack in Patagonia • Southern Ocean sea ice • Baja California vegetation	• Lake levels in high Andes • Vegetation in Somalia • Water levels in Sahel (Chad, Senegal River, Nile) • Caspian Sea water levels • Sumatra smoke • North American snow cover • Snow, lakes in northern California and Nevada • Flooding in Paraná • Sediment plumes from Spanish rivers • Aral′ water levels	• Lake ice, Great Lakes • Lake ice, Baikal • Sea ice, St. Lawrence seaway and North Atlantic Ocean • Snow levels, glaciers in the southern Andes, Alps, Caucasus • General snow blanket over the northern hemisphere (United States, Europe) • Water levels in Caspian, Amu Darya, Turkey, East African lakes, Australian lakes • Vegetation growth in California, Somalia

(Continues)

TABLE 1.1 *(Continued)*

	NASA 2/Mir 21	NASA 3/Mir 22	NASA 4/Mir 22 and 23	NASA 5	NASA 6	NASA 7
Atmospheric events	• Dust storms in Iran, North Africa, Argentina, Andes • Smoke in Mongolia, Amur River • Smoke in Kalahari • smog in western Europe • smog in Taiwan • Ship tracks in clouds	• Tropical storm activity • Thunderstorms • Unique cloud patterns • Atmospheric haze in the eastern United States, India, southern Europe, Mexico • African dust	• Dust in Tibet, Africa, East Asia, India • Smog in southern Europe, Thailand, Santiago • Smoke in Argentina, Mexico, Mongolia, Russia • Unique cloud patterns	• Amazon smoke • Kalahari smoke • South African smog • North Saharan/Algerian dust • Iran–Iraq dust • Nile smoke/smog	• Sumatra smoke • Smog over Italy (Po Valley) • Smoke in central Asia • Smoke in western United States • Smoke over Lake Chad • Dust, Dead Sea Rift • Dust over Aral • Coastal Peru cloud cover	• Smoke in Vietnam • Smoke in Tasmania • Smoke in Mexico and Central America • Dust in Persian Gulf • Dust in Sahara • Dust in Takla Makan and Tian Shan • Smog in Po valley and eastern Mediterranean
Oceans and coastal dynamic features	• Pacific Ocean equatorial fronts • Caspian oil spill • Caspian floods	• Major river mouths: Columbia River, Mississippi/Atchafalaya deltas • Godavari River • Indus River • Gulf of Leyte • Cebu Sea, Philippines • Brazilian coast • Texas and Florida coasts • Bahamian whitings • Northern Chinese plain	• Plankton bloom off Ireland • Major rivers in Europe • Danube delta, Black Sea • Mediterranean coastal region • Indus, and coastal India • Bahamian whitings • Snow melt, Great Lakes • Ice pack, St. Lawrence Seaway	• Sea ice, southern oceans	• Mississippi–Atchafalaya coast • Coastal Texas • Gulf of Venezuela • Persian Gulf dynamics • Red Sea reefs • Caspian Sea flooding • Open ocean swells, wind gusts • Bahamas whitings • Carolina coasts • Sea of Azov • Coastal sediment plumes, Spain	• Caspian flooding • Plankton blooms in South Africa, North Atlantic, Gulf of Mexico, Australia • Deltaic systems: Amazon, Volga, Po, Danube, Godavari, Indus, Colorado, Rhine • Bahamas • Brazilian coast • Texas coast • Australian coast

Volcanoes	• Vesuvius, Italy • Mt. St. Helens, Washington • North Andean volcanic belt • Southern Andes • Damavand, Iran	• Popocatépetl, Mexico • Manam, Papua New Guinea • Galápagos • Mt. St. Helens, Washington • Mt. Rainier, Washington • Taal, Philippines • Elbrus, Russia	• Popocatépetl, Mexico • Galápagos • Mt. St. Helens, Mt. Rainier, Washington • Pinatubo, Philippines • Italian volcanoes • Kamchatkan volcanoes (Karymsky eruption) • Central Asian volcanoes (Elbrus, Damavand)	• Mt. Lascar, Andes, with plume • Virunga Mountains	• Hawaiian haze • Etna, Italy, with plume • Kurile Islands • East Africa maars	• Mt. Etna, Sicily • Pinatubo, Philippines • Kamchatka
Cities/regional targets	• Europe • U.S. Great Lakes • U.S. cities (New York, Boston, Chicago, San Francisco, Seattle) • Patagonia	• Northern Rocky Mountains • Crete • Italy • California • Mexico City	• Europe (Netherlands, Pyrenees, Rhine Valley, Alps, Garonne, Loire, Rhône, Danube, London, Paris, Marseilles, Bordeaux, Salzburg, Munich, Rome, Madrid, Cadiz, Lisbon, Istanbul) • Buenos Aires, Montevideo, Santiago • Detroit, Chicago, Milwaukee, New York City, Boston, Providence • Great Lakes	• South Africa, Tierra del Fuego • Coastal Brazil	• Middle East (Israel, Egypt, Kuwait) • Mediterranean region • Spain • Midwestern United States • California coast, Coastal Texas	• Australia (Perth, Adelaide, Melbourne) • Australian panoramas • Tasmania • New York, Boston, Providence, Philadelphia • London • São Paulo, Rio de Janeiro • Alps, eastern Mediterranean, Netherlands • Central Great Lakes • Central and southern Andes

lasted for more than 10 days in October 1997. Wolf also brought back impressive year-end views of home—the Gulf coast of the United States, including Houston and Galveston.

NASA 7 crew member Andy Thomas completed the two-year Shuttle–*Mir* Earth Science program. During his flight from mid-January 1998 to early June 1998, Thomas, with the *Mir 25* crew of Talgat Musabayev and Nikolai Budarin, completed the documentation of El Niño impacts. NASA 7 imagery included continuation of the time series showing drought conditions in the central Andes and northeastern South America, photos of lush vegetation in California, fires and extensive smoke palls in Central America, and diminished ice cover in the northeastern United States. Thomas also documented short-term events such as a South African plankton bloom, dust storms off Africa and the Takla Makan desert, and more smog over the eastern Mediterranean region. He also brought back extensive coverage of the Australian continent.

1.4.2 Preliminary Science Results

The chapters in this volume summarize some of the imagery highlights taken from the *Mir* between March 1996 and June 1998. Earth observations by crew members over several months and across seasons allow for a significant refinement on the scale and frequency of a variety of dynamic phenomena such as smoke and dust, or unusual precipitation patterns. These data feed into studies on aerosols, dust transport, and land-use changes and recovery, and become part of 1996–1997 baseline data set for validating global parameters such as land cover and water levels. The photos presented in the respective papers illustrate the value of the human observer to recognize and document a wide variety of exciting and quickly changing events on Earth's surface.

Although the photographs are not traditional remotely sensed data, they contain valuable and easily understood information about regional occurrence and duration of hard-to-quantify events. Examples include two global developments in 1997 and 1998: the impacts of the 1997 El Niño (Chapter 5) and worldwide biomass burning (Chapter 7). Both were well documented by astronauts and cosmonauts on *Mir*. An overview of aerosol production and how astronaut photographs contribute to this relatively new field of investigation is provided in Chapter 6. In these discussions we profile the global and regional applications of the data obtained by astronauts and cosmonauts.

We then examine some urban regions of the world through an overview of the changes in North American urban land use over the past 35 years as documented from astronaut–cosmonaut photography (see Chapter 3). An important ancillary database from the Shuttle–*Mir* missions is a set of spectacular photographs of major cities around the world. The imagery has wide and general geographic interest, and individual images will serve as 1996–1997 baselines for quantifying future urban growth.

Coupled with changing land use is the global problem of increasing demands on world water resources. Through images taken from the *Mir*, we have tried to summarize some of the current freshwater issues and highlight the rapid rates of change—from flood to drought—experienced in high-usage watersheds around the world (Chapter 4). This, in turn, leads us to a special collection of papers on the current flooding around the Caspian Sea (Chapters 9 to 19), with a detailed and integrated look at many aspects of the Caspian region. These chapters illustrate the multidisciplinary applications of the photographs in time series for change detection. The data available can support detailed analyses like these for sites around the world.

In Chapter 8 the condition of the *Mir* windows used for imaging the Earth is described. These data, which will help us understand how the space environment affects this part of the optical system, are important for future ISS Earth Sciences data collection. Images from the Shuttle–*Mir* missions are now cataloged and included in the Office of Earth Sciences database. The cataloged data from the Shuttle–*Mir* missions can be accessed on the Web at *http://eol.jsc.nasa.gov*.

1.5 CONCLUSION

Imagery taken from *Mir* on the joint U.S.–Russian Shuttle–*Mir* program is a multifaceted dataset that documents dynamic Earth processes over a two-year period. It provides us with a global perspective on the rhythms and spatial scale of important natural and human-induced events occurring on Earth's surface. Although Visual Observations of Earth by astronauts and cosmonauts has been an ongoing activity for 35 years in both the United States and Russia, much of the information had not been readily available. For the first time, scientists have the global information available to interpret, as a dataset, the thousands of images of Earth brought back from low Earth orbit. The application of imagery to real-world problems has expanded potential as database technologies and global information networks allow universal access to both new and historical imagery, and because individuals and groups can analyze and synthesize global collections of data easily, thus amplifying the power of individual images. These same technologies and networks have also allowed for increasingly sophisticated analysis of the information contained in the images.

The perspective acquired through 35 years of Earth photography by astronauts and cosmonauts allows scientists to identify the important events on Earth and recognize the truly unusual or significant events. Within the context of the database, these new observations from the Shuttle–*Mir* long-duration missions are changing our understanding of the sizes and frequencies of global processes. For example, the biggest recorded fires, some of the biggest recorded dust storms, and the largest oil slick ever documented in the Earth photography database were all captured by *Mir*-based crew members. We believe that we are crossing a new threshold into better understanding of the spatial scales and temporal rhythms of weather patterns, floods, human movements, dust storms, biomass burning, smog production, plankton blooms, and more. If the experiences of the Shuttle–*Mir* astronauts and cosmonauts are typical, Earth observations by crew members on the International Space Station (ISS) will greatly improve both our database and our understanding of processes and changes on Earth.

ACKNOWLEDGMENTS

The chapters in this volume and these views of the world would not have been possible without the hard work and persistence of many people, especially the astronaut and cosmonaut crews who collected the data: Yuri Usachov and Yuri Onufrienko (*Mir* 21), Shannon Lucid (NASA 2), John Blaha (NASA 3), Valery Korzun and Alexander Kalery (*Mir* 22), Jerry Linenger (NASA 4), Vasili Tsibliyev and Alexander Lazutkin (*Mir* 23), Michael Foale (NASA 5), Anatoly Soloviev and Pavel Vinogradov (*Mir* 24), David Wolf (NASA 6), Andy Thomas (NASA 7), Talgat Musabayev and Nikolai Budarin (*Mir* 25). Support from terra firma included Ivan Firsov (RSC Energia), the Shuttle–*Mir* Mission Science and Operations Support teams, and the Russian support teams. These people are also gratefully acknowledged. Marv Glasser compiled Figure 1.3.

REFERENCES

Amsbury, D. L. 1989. United States manned observations of Earth before the Space Shuttle. *Geocarto International,* 4(1):7–14.

Anonymous. 1977. *Skylab Explores the Earth.* National Aeronautics and Space Administration Special Publication 380, 517 pp.

El-Baz, F., and Warner, D. M. 1979. *Apollo–Soyuz Test Project, Summary Science Report.* Vol. II: *Earth Observations and Photography.* National Aeronautics and Space Administration Special Publication 412, 692 pp.

Ewing, G. C. 1965. *Oceanography from Space.* Ref. 65-10. Woods Hole, MA: Woods Hole Oceanographic Institution, 469 pp.

Foster, N. G., and Smistad, O. 1967. *Gemini Experiments Program Summary.* National Aeronautics and Space Administration Special Publication 138, pp. 221–230.

Helfert, M., and Lulla, K. P. 1989. Human-directed observations of Earth from space. *Geocarto International Special Issue,* 4(1):80 pp.

Koval, A., and Desinov, L. 1987. *Space Flights Serve Life on Earth.* Moscow: Progress Publishers, 283 pp.

Lowman, P. D. 1985. Geology from Space: a brief history of orbital remote sensing. In Drake, E. T., and Jordan, W. M., eds., *Geologists and Ideas: A History of North American Geology,* Centennial Special Vol. 1. Boulder, CO: Geological Society of America, pp. 481–519.

Lowman, P. D., and Teidemann, H. A. 1971. *Terrain Photography from Gemini Spacecraft: Final Geologic Report.* Report X-644-71-15. Greenbelt, MD: Goddard Space Flight Center, 75 pp.

Lulla, K., Evans, C., Amsbury, D., Wilkinson, J., Willis, K., Caruana, J., O'Neill, C., Runco, S., McLaughlin, D., Gaunce, M., McKay, M. F., and Trenchard, M. 1996. The Space Shuttle Earth observations photography database: an underutilized resource for global environmental sciences. *Environmental Geosciences,* 3(1):40–44.

Chapter

2

Russian Visual Observations of Earth: Historical Perspective

Nikita F. Glazovskiy and Lev V. Dessinov

Institute of Geography
Russian Academy of Sciences
Moscow, Russia

ABSTRACT

The Russian experiment, Visual Observations (photography of Earth by cosmonauts) started with the earliest Soviet space flights. This chapter provides an historical overview of the Russian campaign to document the Earth's environment from space. Details of the program philosophy, the evolution of the hardware and spacecraft capabilities, and some of the key participants involved in shaping the experiment are described.

2.1 INTRODUCTION

Visual observations of Earth's environment by humans in space began on April 12, 1961, when Earth's first cosmonaut, Yuri Gagarin, saw our planet through the window of the *Vostok* spacecraft.

In the history of Russian cosmonautics, experiments in visual observation and photography of Earth's surface consisted of five stages. The first stage, from 1961 to 1974, was one of enthusiastic impressions. In those years, the first information on the appearance of Earth was obtained, including detailed cosmonaut accounts and thousands of photographs. It is important to note that in those years, visual observations were essentially spontaneous, without any kind of plan, and with minimal input from earthbound scientists. It is problematic to label the early photography of Earth as experiments because the important and unique results were the fruit of the creativity of the cosmonauts themselves.

The Russian cosmonauts' accounts and photographs of Earth were analyzed in detail at scientific centers in the USSR and abroad. The theoretical possibilities for orbital investigation of Earth's surface were confirmed, and the fundamental advantages of visual observations from orbit became evident. These included (1) the capability for global, real-time observations of the planet, (2) a characterization of the general appearance of Earth's surface, (3) the study of geographically inaccessible territories and regional investigations of features and events that may cross political borders, (4) comparative analyses of similar phenomena and features occurring in different regions, and (5) the documentation of large, dynamic events (both natural and human-induced).

The second stage of visual observations of Earth lasted from 1974 to 1976. By early 1974, the need to implement a program of visual and instrumental observations with specific objectives had become evident. Such an initiative was developed by a group of scientists, including Doctor of Geography and Cosmonautics Lev Desinov (manager), Doctor of Geography Igor Abrosimov, Doctor of Geology Vladimir Kozlov, and Doctor of Cosmonautics Aleksandr Koval.

The project received support but at first was implemented unofficially. The people who made the decisions were guided by the principle of, "Let's see what comes of it." However, a most-favored status was given to the program at the Yu. Gagarin Cosmonaut Training Center, whose chief, General and Cosmonaut Georgiy Beregovoy, personally supported the project. On his recommendation, Air Force Colonel and Honored Pilot-Navigator of the USSR (Certificate No. 1) Nikolay Zatsepa set about organizing the aerial visual-training flights of cosmonauts and scientist-instructors. Another important player was RSC-Energia engineer Ivan Firsov, who saw to the photographic recording of flights over a 37-year period.

The second-stage space flights were conducted by the Salyut 3 crew Pavel Popovich and Yuri Artyukhin (June 25, 1974 to July 19, 1974), the two crews of the Salyut 4 orbital station Aleksey Gubarev and Georgiy Grechko (January 11, 1975 to February 9, 1975) and Petr Klimuk and Vitaliy Sevastyanov (May 24, 1975 to July 26, 1975), and the Salyut 5 crews Boris Volynov and Vitaliy Zhelobov (July 6, 1976 to August 24, 1976) and Viktor Gorbatko and Yuri Glazkov (February 7, 1977 to February 25, 1977).

The chief technology of the 1974 to 1976 experiments was portable and fixed binoculars. The portable binoculars had 12× magnification; the fixed binoculars—called the optical viewer—had a mass of around 100 kg and enabled observation of natural objects with a magnification up to full-image blurring. One remarkable feature of this instrument was the ability to stop the "rotation of the Earth" and view a fixed picture. The binoculars also had a photographic attachment.

Another important feature of the experiment was the fact that all observations were made in a strict orbital attitude. An electromechanical system for deviation damping was installed on the Salyut 3 and 5 stations for this. This feature made it possible to document the results of observations using fixed cameras rigidly secured to the station floor when necessary. The KATE-140 camera was the one used most often. It was equipped with an objective lens with a focal length of 140 mm. With an 18 × 18 cm frame format, the camera could take pictures of a terrain area of approximately 440 × 440 km. Conventional hand-held cameras with medium film formats (56 × 56 mm) and 24 × 36 mm frame format (Praktika-E, produced in the German Democratic Republic) were also used extensively.

The main stage in the Russian visual observation program was from 1977 to 1985, including observations from Salyut 6 and 7 (Table 2.1). During this period, dozens of scientific centers in the USSR and socialist countries were involved in the program. Each crew, regardless of mission duration, underwent natural science training and had their own visual observation log on board with an explanation of the essence of each experiment. The cosmonauts worked closely with scientists during and after the flight and the principal investigators of the visual observation experiments had access to the cosmonauts almost immediately after their return from orbit.

The Salyut 6 and 7 experiments included more complex observational tasks. Observations were made generally in unoriented station flight and were frequently oblique or obscured by solar panels. Another important component of these experiments was the long duration of the main missions, with a maximum of 237 days. Sometimes by the end of a mission, the cosmonauts were essentially Earth science teachers, and they performed a continuous function as mentors for participants of short missions, including international ones. The observations program was expanded through scientists from those countries whose cosmonauts were on board the station.

The *Mir* space station, launched in 1986, brought the fourth stage of Russian earth science observations from space. From 1986 to 1993, the visual experiments were continued on *Mir,* but they were spontaneous in nature, since financing had been sharply reduced. The fifth stage, U.S.–Russian cooperation in Earth science from space, began with joint Space Shuttle flights in 1994 and seven joint *Mir* missions between 1996 and 1998. The results from this cooperation are highlighted in this volume.

TABLE 2.1   Flights on the Salyut 6 and Salyut 7 Orbital Stations

Cosmonauts	Time of Flight	Mission Duration (days)
Salyut 6		673 total
Yuri Romanenko, Georgiy Grechko	December 10, 1977–March 16, 1978	96
Vladimir Kovalenok, Aleksandr Ivanchenkov	June 15, 1978–November 2, 1978	140
Vladimir Lyakhov, Valeriy Ryumin	February 25, 1979–August 19, 1979	175
Leonid Popov, Valeriy Ryumin	April 9, 1980–October 11, 1980	186
Vladimir Kovalenok, Viktor Savinykh	March 12, 1981–May 26, 1981	76
Salyut 7		766 total
Anatoliy Berezovoy, Valentin Lebedev	May 13, 1982–December 10, 1982	211
Vladimir Lyakhov, Aleksandr Aleksandrov	June 27, 1983–November 23, 1983	150
Leonid Kizim, Vladimir Solovyev, Oleg Atkov	February 8, 1984–December 2, 1984	237
Vladimir Dzhanibekov, first with Viktor Savinykh and then with Vladimir Vasyutin and Aleksandr Volkov	June 6, 1985–November 21, 1985	168

2.2 DETAILED LOOK AT SALYUT 6: VISUAL AND INSTRUMENTAL OBSERVATIONS OF THE ENVIRONMENT FROM LONG-DURATION MISSIONS

In looking back, one must admit that the part of the program that was carried out during the flight of the Salyut 6 station was the most fruitful. Any human space flight, even a short one, presupposes the study of geographic features by means of visual and instrumental observations. Experience shows that visual observations are a significant complement to photographs, as they make it possible to ascertain many features of the outward appearance of Earth's surface. This result is achieved by the exceptional sophistication of the human eye and also by the combination of high visual selectivity and logical analysis of an observed image. No single instrument in existence is capable of fully recording what a cosmonaut sees. Furthermore, requirements can be developed for the design of new instruments and their operating procedures determined on the basis of such observations.

Many natural science tasks can be carried out solely by examination from space. For example, what other method could quickly detect icebergs in the ocean and track their paths effectively? What other method could register the appearance of dust storms in the deserts?

For crews of the Salyut 6 orbital station, a program of visual observation of Earth was developed based on synthesis of the information of cosmonauts who had been in orbit from 1961 to 1977. In addition to the traditional duties of controlling the remote-sensing equipment, this program asked the cosmonauts to select objects of investigation and their observation times according to the weather situation, lighting conditions, and many other factors.

2.2.1 Training

To support this program, it was deemed necessary to help them look at Earth through the eyes of a natural scientist. Based on the assumption that several long-duration mission crews would be on the Salyut 6 station, specialists decided that during the 96-day flight of the first crew they would concentrate on developing procedures for space visual observations so as to achieve the maximum effectiveness in subsequent investigation stages.

An extensive training curriculum was developed for the cosmonauts, including the international partners. All phases of the earth science training were carried out in Star City at the Yu. A. Gagarin Cosmonaut Training Center in a natural science procedures laboratory equipped with technical training systems. At individual workplaces, students have the opportunity for independent study of Earth science fundamentals. Each workplace is equipped with a monitor to examine space photos, a tape recorder, and other accessories. In addition to individual classes, experts from various Earth sciences and Star City methodologists described Earth landscapes to the cosmonauts, including the distinguishing signs of various natural formations, types of pollution, and all sorts of basic phenomena. Theoretical knowledge was reinforced in training and demonstration flights in the Tu-134 aircraft laboratory. The objects selected for aerial visual observation were regions that the cosmonauts would be studying from orbit, along with scientific test ranges.

Cosmonauts also studied the onboard environmental visual observation logs in which all the scientists' specific assignments and recommendations were provided in short form. These assignments were illustrated with maps and diagrams and often also by photos of the objects to be observed. They also contained tables for recording the results of observations and recommendations on photography.

2.2.2 Operational Results

The results of the work of cosmonauts Yu. V. Romanenko and G. M. Grechko were outstanding. First, the cosmonauts compared several types of maps against the actual underlying surface pattern and found that maps with a scale of 1:2,500,000 were most suitable for naked-eye observation of Earth from an altitude of 350 km and more. They offered recommendations to improve such maps. For example, they suggested that a specific geographical region be shown in maps in the color that distinguished it from other territories. Thus, South American looks dark green from orbit, Africa looks brown, Australia is bright yellow, Tibet red, and so on.

The cosmonauts tested several types of binoculars and opted for 12× magnification. After testing various color atlases, they concluded that it was necessary to send up a colorimeter for objective quantitative color measurement (Vasyutin and Tishchenko, 1989). One specific feature of the Salyut 6 station was its flight in drift mode, which was accompanied by random wobbling and rotation about the center of mass. To record Earth's surface using a camera or spectrometer, a station active stabilization mode was generally used, with the help of low-thrust jets. In this case, the longitudinal axis of the Salyut–Soyuz complex was kept perpendicular to the local vertical, for which purpose an economical but nonetheless propellant-consuming attitude control and stabilization system was used. During the work of photographing Earth with fixed cameras, both crew members were kept busy running the photographic and spectroscopic hardware and monitoring stabilization of the complex. They also recorded results in the flight data files and prepared reports to the mission control center.

The crew had real opportunities for visual and instrumental observations of Earth only when the station was in drift mode. However, in a number of cases, upon approaching the observation region, when the cosmonauts took their places at the windows after preparing binoculars and cameras, the station would rotate so that its solar arrays blocked the view of Earth.

The crew very quickly noted an obvious pattern: The station always wobbled around its transverse axes, on the average maintaining an up-or-down orientation of the solar arrays, and never moved quickly from one type of attitude to another. This fact suggested to the cosmonauts that they establish a station attitude with the longitudinal axis along the local vertical and impart an angular velocity to it on the orbital plane equal to the velocity of orbital rotation about the Earth, damping the two other orthogonal components of the angular velocity. In other words, the cosmonauts decided to implement in practice the theoretically known possibility for gravitational stabilization mode. Since there were no special damping devices on the station to damp the natural oscillations in the orbital coordinate system, the required initial conditions were produced using the manual controls of the attitude control microjets and the viewer.

The effect from such station rotational motion dynamics proved to be excellent. The image of Earth in the viewer became almost fixed, and the wobble about a given position was minimal. There were periods when the station was in a state of gravitational stabilization for up to seven days, from one dynamic mode to another. The crew obtained the opportunity for round-the-clock observations from the station transfer compartment, which was equipped with five windows. After establishing a sort of environmental visual and instrumental observation post, the crew transmitted several hundred messages regarding objects of investigations, accompanying many of them with photos using the onboard portable camera.

Nearly every day of the 96-day flight of Yu. V. Romanenko and G. M. Grechko brought interesting and important messages that attracted close scientific attention. For example, one cosmonaut reported, "Detected oil slick off the southern coast of Africa, length is around one hundred kilometers. It is moving from the tanker collision site toward the southwest and is slightly eroded by the current." This message was especially important to experts—the origin of the slick was well known, so using it as an example, one could study how such pollution looks from orbital altitude, and in future observations this phenomenon would not be confused with similar ones.

One should also note the important role of reference objects already known to the cosmonauts from their training and demonstration flights in aircraft. When detecting linear geological structures, the cosmonauts compared their observations against models of the Talaso–Fergan fracture in central Asia and the Dzhazlair–Nayman and Main Karatausk faults in Kazakhstan. When studying hydrological features of arid regions, the references were usually indicators associated with the development of vegetation, various types of sands, or ancient river beds in Turkmenistan and Uzbekistan. Observing familiar natural ranges on the territory of the southern USSR from orbit, the cosmonauts offered recommendations to improve the theoretical training and programs of preliminary aerial visual observations for crews of subsequent missions.

During their 96 days in orbit, an evaluation was made of the influence of weightlessness on the effectiveness of visual observation of Earth. It was noted that visual acuity decreases slightly during the first days of flight but then remains practically unchanged after adaptation to weightlessness. The overall quality of scientific information on the natural environment improved as the observers acquired experience in visual examination of Earth. This was taken into account in planning visual and instrumental observation programs for crews of short-term visiting missions, including international crews. The long-duration crews were required to conduct a preliminary examination of observation targets to evaluate optimal observation conditions and to acquire experience in rapid detection using reference points readily visible from orbital altitude. The main crew assisted the visiting cosmonauts in performing their environmental investigation programs.

Another organizational form worth noting was the operational message to the crew, prepared in the form of an express report that summarized performance of visual and instrumental experiments and the first results of their analysis. This report was sent to the Salyut 6 station for the first time with the first visiting mission. It contained a digest of completed investigation stages and focused cosmonaut attention on the main objectives of the upcoming period.

At the end of the 96-day flight of the first main mission, it was possible to develop recommendations regarding the environmental visual investigation procedure, which included the following operations:

1. Study the flight mission using the flight data files and operational supplements to it; prepare the observation and recording systems.
2. Check station stability relative to its orbital motion velocity vector, and perform dynamic operations if necessary.
3. Detect the observation object in the terrain using available cartographic materials and space photos on board and the characteristic reference points.
4. Distinguish and observe the specific features of the object in accordance with the assignment.
5. Photograph observation regions using the onboard hand-held camera.
6. Record the object on the map, diagram, or photo in the onboard log; record observation and photographic results (fill out the corresponding tables) and sketches if necessary.
7. Report to Mission Control Center, and consult with the experts and developers of the experiments.
8. Perform additional study or photography of the object when new expert recommendations were received.

During the 140-day work of cosmonauts V. V. Kovalenok and A. S. Ivanchenkov on Salyut 6, environmental visual and instrumental investigations became even more extensive. The second crew continued the experiments using the experience of their predecessors. For example, they obtained unique results on the relation between ocean currents and the meteorological situation over the ocean, on the conditions for observation of submarine terrain features, and on mountain glaciers.

The cosmonauts also participated in an experiment to determine the actual resolution of the eye when observing details of the natural landscape. To do this, they selected several glaciers in the Pamir mountains as a test section. They included space photos of the Pamir range and ground photos of glaciers of this region in the flight data file. The Bivachnyy glacier (Figure 2.1) was conditionally divided into 40 square sections. The crew was asked to describe those elements of the external appearance of the glacier which they observed with the naked eye and with their binoculars. At they same time, a helicopter patrol was carried out over the Pamir mountains, during which the cosmonauts' messages were checked and evaluated.

The experiment showed that when there is a high contrast of landscape elements, the human eye without binoculars can distinguish elongated natural formations just 15 to 20 m wide from a distance of 350 km. This is a few seconds of arc. Earlier, it was thought that only natural objects with angular dimensions of more than 1 minute could be distinguished from such an altitude.

Another important result of investigations by cosmonauts of the second mission was their report on evaluation of the effective observation time of several dozen types of natural formations. For example, it was found that the drifts of major fires could be tracked for 4 to 5 minutes, but mountain pastures in narrow valleys for no more than 10 to 15 seconds.

Using experience from the first two crews, cosmonauts V. A. Lyakhov and V. V. Ryumin of the third long-duration mission focused their primary attention on studying ocean waters. They held 53 communication sessions with oceanographers and

Figure 2.1 Space photo of the Pamir Mountains from Russian Resurs F1 satellite. (*A*) Regional view of Fedchenko glacier (the long, north-trending glacier), and Bivachnyy glacier (marked with an arrow). Communism Peak is the high, snow-covered peak near the left edge and above the center of the image. (*B*) Detailed view of the area defined by the box; also, Bivachnyy (arrow) and Fedchenko glaciers.

fisheries experts. The total length of cosmonaut messages about studies of ocean waters was around 35 hours. In 175 days of flight, this crew successfully photographed the ocean through the side windows more than 300 times.

For 185 days, the crew of the fourth main mission, cosmonauts L. I. Popov and V. V. Ryumin, observed cloud formations, volcanic eruptions, and linear and circular geological structures, continued their oceanographic investigations, and transmitted information on mountain glaciers. The crew sent down 680 photos taken with portable cameras.

During the time crews were aboard the Salyut 6 orbital station (596 days), the KATE and МКФ-6М cameras were used to photograph Earth landscapes roughly 9000 times. These cameras withstood a three-year test of weightlessness and were recommended by specialists for use on subsequent orbital stations. As for the portable cameras, their successful use of color reversal films was noted. The high-resolution films were relatively resistant to cosmic radiation, less sensitive than others to errors in exposure selection, and most fully conveyed the color features of the landscape.

2.3 CONCLUSION

Through the attention and efforts of the early cosmonauts and Earth scientists working with them, visual observations of Earth from space have continued for more than 30 years. The style and procedures established by the early Russian crews proved to be operationally efficient and robust. Visual observations of Earth from space continues today in much the same way and has yielded thousands of images for scientists around the world.

REFERENCES

Editor's Note: The following English-language publications provide additional information on the history of Soviet spacecraft and Russian visual observations of Earth.

Hart, D. 1987. *The Encyclopedia of Soviet Spacecraft.* New York: Exeter Books, 192 pp.
Koval, A., and Desinov, L. 1987. *Space Flights Serve Life on Earth.* Moscow: Progress Publishers, 286 pp.
Vasyutin, V. V. and Tishchenko, A. A. 1989. *Space Coloristics.* Scientific American, July, p. 84–90.

Chapter

3

Twenty-Eight Years of Urban Growth in North America Quantified by Analysis of Photographs from Apollo, Skylab, and Shuttle–*Mir*

Julie A. Robinson, Brett McRay, and Kamlesh P. Lulla

Office of Earth Sciences
NASA Johnson Space Center
Houston, Texas USA

ABSTRACT

Rapid land-use change around urban areas has become a sensitive environmental and political issue in the late twentieth century. We quantified growth of North American cities using photographs taken by astronauts from low Earth orbit. The study included a set of six North American cities that have expanded over the last 30 years, including Vancouver (British Columbia), Chicago, San Francisco Bay area, Dallas/Fort Worth, Las Vegas, and Mexico City. We identified baseline photographs of each city from Apollo 9 (1969) or Skylab (1973–1974) and paired them to photographs taken during the Shuttle–*Mir* program (1996–1998).

The photographs were digitized, registered to each other, and resampled to a uniform per-pixel scale. We used visual photointerpretation to delimit the boundary of built-up area, and manually digitized the boundary to the scale of our registered images. The approximate scale of a pixel in the registered images was calculated by measuring the length of an airport runway in each scene. We expected that our built-up area estimation would fall between the city-limits and urban-agglomeration estimates of the United Nations (UN), and this was observed for Chicago, Dallas, Las Vegas, and Vancouver. The sprawling area we observed during photointerpretation of the San Francisco Bay area led to a built-up area estimate that far surpassed that for the UN urban agglomeration area.

During the period we studied, Dallas, Mexico City, Chicago, and Las Vegas expanded at rates of a similar order of magnitude (18.3 to 53.8 km^2/yr), while Vancouver expanded at a rate of 4.9 km^2/yr and the San Francisco Bay area expanded at

136.1 km²/yr. Some cities increased in built-up area disproportionately relative to their increase in population. For all cities other than Las Vegas, the expansion rate of built-up area exceeded the expansion of population, indicating urban sprawl. Las Vegas was an exception to this trend because of the number of discontinuous communities included in the urban agglomeration as redefined by the U.S. Census in 1994. Although the annual areal expansion of Las Vegas was comparable to other cities, its small initial size means that it tripled in area from 1973 through 1996.

These results illustrate the potential for further quantitative analysis of urban change using astronaut photographs. Our methods provide a less effort-intensive alternative to standard urban remote sensing techniques that would be appropriate for quantifying large-scale urban impacts on the landscape. They also are appropriate for research questions requiring analysis of numerous cities.

3.1 INTRODUCTION

The United Nations (1991) predicted that by the year 2000, more than 50% of the world's population would live in urban areas and that this would increase to 65% by 2025. Concomitant with this urbanization has been linkage of urban areas into agglomerations (megacities) and increasing urban sprawl. Such changes are of interest to social scientists and urban planners (e.g., Alig and Healy, 1987; Geddes, 1997). In addition, urbanization is an important component of change in the land, air, water, and biota (Rapoport et al., 1983; da Costa, 1991; Meyer, 1996; Cicero-Fernández, 1997; Spengler and Ford, 1997). Urbanization is linked to loss of biodiversity because "urban areas are effectively synonymous with ecosystem disruption and the erosion of biological diversity" (Murphy, 1988:72). Recognizing the importance of ecological aspects of urban change, a framework for ecosystem studies along urban–rural gradients was proposed by McDonnell and Pickett (1990). The National Science Foundation recently announced a research initiative soliciting proposals for studies of change in urban environments (Brown, 1997; National Science Foundation, 1998).

Quantification of urban change requires time-series data at a suitable level of detail plus an analysis method that optimizes level of effort. Ground surveys of urban change are laborious and expensive. Acquiring the numerous aerial photos required to monitor urban change on a regular basis can also be extremely costly. Data collected from remote sensors in Earth orbit provide a much less expensive alternative at the potential cost of detail (Paulsson, 1992).

3.2 REMOTE SENSING FOR URBAN ANALYSIS

Some of the earliest remotely sensed images of entire cities came from astronaut photography. Early in urban remote sensing research, investigators evaluated satellite photographs as an alternative to ground surveys for estimating population sizes of cities. Nordbeck (1965) proposed that a relationship between the urban area (A) of a settlement and its population (P) would take the form $A = aP^b$, where a and b are empirically derived constants. Wellar (1969) applied this model to Gulf coast communities using built-up areas estimated from Gemini space photography. Ogrosky

(1975) performed visual interpretation of Skylab photos of small U.S. cities and found that population was best predicted by urbanized area (regression $R = 0.982$). It appeared that the model was less applicable to larger cities [e.g., Chinese cities of >2,500,000 people (Welch, 1980); 40 cities in the U.S. Tennessee Valley 2500 to 99,999 in population (Holz et al., 1969)]. One approach to failures of the model was to attempt to make determination of the urban area boundaries more "objective" through the use of digital image analysis (Anderson and Anderson, 1973). The hope that satellite data could provide an alternative to ground population surveys subsided as researchers realized that variations in settlement patterns required different resolutions to delimit urban land-use types (Welch, 1982; Jensen, 1983).

Urban remote sensing research began to emphasize accurate land-use mapping at high levels of detail (Level III land-use classes, i.e., single-family residential, multi-family residential, etc.) (Anderson et al., 1976, Lins, 1976). Detailed change detection (i.e., of change among Level II and III land-use classes) generally has not been successful with unmanned satellite data alone (review by Jensen, 1986; see also Christenson et al., 1979; Jensen, 1979, Friedman, 1980; Martin et al., 1988). Subsequent work has emphasized increasing the detail of information obtained by improving image analysis techniques (e.g., Howarth and Boasson, 1983; Hill and Hostert, 1996), sensor resolution (e.g., Welch, 1982; Toll, 1985; Ehlers et al., 1990), use of ancillary data (e.g. Barnsley et al., 1989; Møller-Jensen, 1990; Sadler and Barnsley, 1990; Kam, 1995), and change detection algorithms (Stow et al., 1990). Much of this emphasis probably has been driven by the needs of municipalities for detailed updated urban land-use information (Paulsson, 1992).

Although most urban remote-sensing studies since the mid-1970s have used data from robotic orbital scanners (such as Landsat and SPOT), the dramatic views that astronauts see of many urban areas have motivated continued and frequent astronaut photography. Astronaut photographs are a uniform and robust data source that covers the broadest possible range of dates [beginning with photos taken during the Gemini program (1965 to 1966) and continuing through the present]. In contrast, Landsat MSS became available in 1972, Landsat TM in 1982, and SPOT in 1986 (Campbell, 1996). Once photographic data are converted to digital form and resampled to scale, the resulting pixel sizes can be compared to pixel sizes for other remote-sensing data. As first reported for Skylab photographs (Lins, 1976), the resolution and classification potential of the photographs can exceed that of more widely used satellite scanner data. In practice, we have been able to resolve objects as small as 8 m on the best astronaut photographs of cities (J. A. Robinson, D. Liddle, and C. A. Evans, unpublished data). A detailed treatment of resolution issues in these photographs is in preparation.

3.3 OBJECTIVES

The early approach of measuring urban area boundaries is still applicable to questions of urbanization at scales larger than a single municipality. For example, many studies of the ecological impacts of urban sprawl do not require details on the conversion of land within Level III land-use types (e.g., Castiaux et al., 1991; da Costa, 1991; Mishra et al., 1994). Simple methods to delineate gross urban area boundaries may be useful in regional or global studies of the environmental impacts of urbanization. Advances in image-processing technology could be used to improve our ability to measure urbanization boundaries without requiring expensive detailed studies of land-use change within the city.

As a test of this concept, we developed a method of estimating overall urban area change that could readily be applied to a large number of cities in comparative analyses. To meet this objective we believed that (1) the data used must be in the public domain without expensive acquisition costs, (2) the data used should cover as long a period as possible, and (3) image analysis techniques should be kept as simple as possible. We were not interested in supplanting or augmenting current approaches to estimating urban populations per se, but rather, to provide an alternative method for estimating the extent of urban areas. A secondary objective was to illustrate the use of this approach in understanding patterns of urbanization by comparing the built-up area measured for each urban region to independent estimates of populations and urban areas.

3.4 METHODS

We selected a set of North American cities that had been photographed during Apollo 9 (1969) or the three Skylab missions (1972 to 1974), and again from the *Mir* space station during the Russian–U.S. joint Shuttle–*Mir* program (1996 to 1998). These photographs are in the public domain and available at cost of reproduction (see Office of Earth Sciences, 1998). Data selected were not generally the best photographs ever taken of each city (e.g., "Earth from Space—Search (Cities)" undated; Au, 1993). However, they were of suitable quality for remote sensing and covered similar time frames [Figures 3.1 (see the color insert) and 3.2 to 3.4]. Photos were identified by searching the publicly accessible database of NASA Earth-observing photos (Earth Science Branch, 1997), compiling a list of photos from suitable missions, and visually examining the film to select photographs. Selection criteria included minimal cloud cover, good focus, proper exposure, and inclusion of the subject city with surrounding area. We excluded photographs with snow cover that might alter the visual appearance of city boundaries but did not consider other potential seasonal differences.

We digitized third-generation film at 1200 and 2400 pixels per inch (ppi) using a flat-bed scanner (Arcus II, Agfa-Gevaert N.V., Mortsel, Belgium) and Photoshop software (version 4.0, Adobe Systems Inc., Mountain View, California) and imported the resulting three-band image into ERDAS Imagine software (version 8.3, ERDAS, Inc., Atlanta, Georgia; configured on a Silicon Graphics Indigo2, SGI, Mountain View, California). For each city we selected the most nearly nadir view (usually the best resolution) as the reference image. We estimated angle of rotation for north orientation comparing linear features in the photograph to their representation on 1:1,000,000 Aircraft Operational Navigational Charts (U.S. Defense Mapping Agency). Although more detailed maps (e.g., U.S. Geological Survey topographic maps) could have been used for areas in the United States, we chose the navigational maps because they provide uniform worldwide coverage. The reference image was rotated and resampled using an affine transformation with a bilinear convolution algorithm (ERDAS, 1997).

We registered the image with the more oblique view to the more nearly nadir-looking image by developing a first-order polynomial model with a minimum of 20

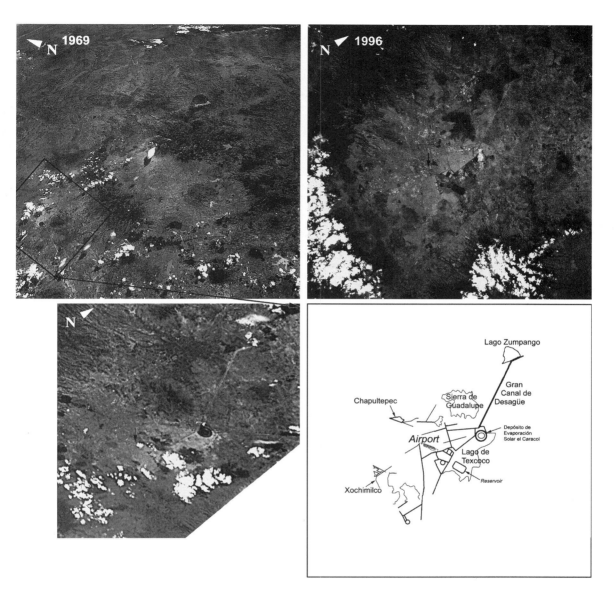

Figure 3.2 Photographs of Mexico City in 1969 and 1996 (NASA photograph AS9-19-3011 and NM22-741-54B). These two images differed greatly in scale and look angle. It was still possible to georeference the images to one another. The inset shows detail from the 1969 image that has been rotated to match the orientation of the 1996 image. A sketch map illustrating the major canals and other distinctive features visible in the images. Small arrows indicate Licenciado Benito Juarez International Airport used for scale.

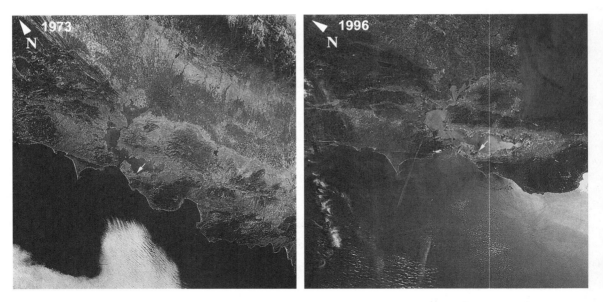

Figure 3.3 San Francisco Bay area in 1973 and 1996 (NASA photograph SL3-121-2374 and NM21-741-15A). These photographs were chosen over sets at more detailed scales to avoid the need to mosaic multiple images of such a large urban region. Small arrows indicate San Francisco International Airport used for scale. See also Figure 4.9.

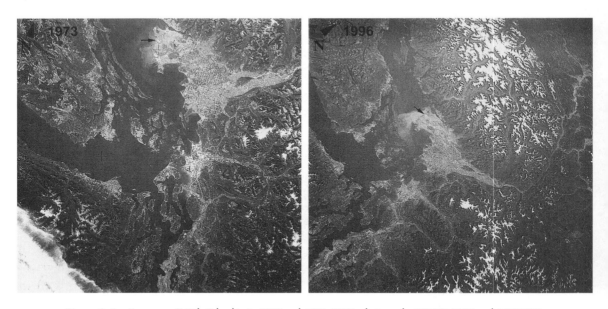

Figure 3.4 Vancouver, British Columbia in 1973 and 1996 (NASA photographs SL3-122-2512 and NM22-703-120). The original 1996 image was significantly underexposed, but detail for analysis was retrieved by darkroom lightening of prints and digital lightening of the digitized image. Small arrows indicate Vancouver International Airport used for scale.

tie points that were visible in both photographs. This was equivalent to transforming the raster grid of the lower-resolution image data to that of the higher-resolution raster grid (cf. Stow et al., 1990). We first chose three ground-control points uniformly arrayed around the approximate urban area. Additional ground-control points were selected to be uniformly distributed in and around the city, with root-mean-square (RMS) error and total RMS error determined incrementally as each point was added (ERDAS, 1997). Based on our experience in registering multitemporal images to each other, our accuracy goals for a minimum total of 20 control points were that (1) no single point should exceed a RMS of five pixels, and (2) total RMS error after all points were added should also be less than five. However, we did recognize that variations in image quality and scale might make this difficult for some transformations.

We compared our ability to select and interpret tie points in the images digitized at 1200 and 2400 ppi and found no difference in RMS errors in the registration. Therefore, we used the 1200-ppi image to speed image processing. We then rotated and resampled the registered second image to north orientation (using an affine transformation with bilinear convolution algorithm). A subset was defined on each image that included 40 to 50% of image space around the apparent area of the city. We estimated the approximate scale for each pair of city images by measuring the distance along a clearly identifiable airport runway. The length of an airport runway in image pixels combined with the known length of the runway on the ground (Aircraft Owners and Pilots Association, 1997) provided the scale.

In pilot studies we attempted simple classifications of urban areas based on grayness. Results included an unacceptable level of misclassification. We could have significantly improved automated remote sensing results by including other factors, such as texture, that are commonly used in the literature, but this would not have met our stated objective of keeping the methods as streamlined and minimal as possible. Therefore, we chose to use visual photointerpretation of 8 by 8 photographs of the original image (cf. Holz et al., 1969) to delimit built-up area boundaries. Throughout this chapter we use the term *built-up area* to refer to this area identified by photointerpretation, and *urban area* as a general descriptive term for an area typified by urban land uses. We then digitized these boundaries by hand and counted the number of pixels in each area.

One side effect of the visual photointerpretation was that boundaries were smooth; noncontiguous built-up areas were not included, and any "islands" within the built-up areas were included within the built-up area polygon. The combination of visual photointerpretation with computer image processing also meant that error in determining city boundaries was due only to human error in interpretation and digitizing. Any introduction of error due to mixed pixels occurred as the boundary was hand-digitized and not from the digital resampling.

We used population census data drawn from compilations in the UN's *Demographic Yearbook* as a rough test of the order of magnitude of our measurements (Table 3.1). In a special feature of the 1995 yearbook, areal estimates were made for cities and urban agglomerations, and we also compared our results to these area measurements. We expected that our built-up area estimates should be larger than tabulated values for cities proper, but smaller than tabulated values for urban agglomerations, which generally included multiple disjunct urban centers. We relied on UN compilations rather than on more detailed (and presumably more accurate)

TABLE 3.1 Reference Data for Cities Used in This Study

Urban Area	Image Year	Population[a]		Demographic Yearbook[b]	Built-up Area[a] (km²)	
		City Proper	Urban Agglomeration		City Proper	Urban Agglomeration
Chicago[c]	1973	3,457,237 (1973)	7,612,314 (1970)	1975		
	1996	2,845,999 (1994)	8,526,804 (1994)	1995	589 (1990)	2,676 (1990)
Dallas[d]	1969	1,237,877 (1970)	2,318,036 (1970)	1974		
	1996	1,761,566 (1994)	4,362,483 (1994)	1995	1,856 (1990)	3,915 (1990)
Las Vegas[e]	1973	144,333 (1973)	273,288 (1970)	1975		
	1996	327,878 (1994)	1,076,267 (1994)	1995	216 (1990)	15,201 (1990)
Mexico City	1969	6,874,165 (1970)	8,589,630 (1970)	1973		
	1996	8,235,744 (1990)	15,047,685 (1990)	1995		
San Francisco[f]	1973	1,784,822 (1973)	4,628,199 (1970)	1975		
	1996	2,146,932 (1994)	6,513,322 (1994)	1995	902 (1990)	2,845 (1990)
Vancouver	1973	426,260 (1971)	1,137,000 (1974)	1975		
	1996	471,844 (1991)	1,602,502 (1991)	1995	113 (1991)	2,786 (1991)
			1,774,672 (1994)	1995		

[a]Number in parentheses indicates year data collected.
[b]UN Department of Economic and Social Affairs, New York.
[c]Greater Chicago city proper in 1973 is the sum of Chicago, Gary, and Hammond (but not Kenosha), and in 1994 the sum of Chicago and Gary. Urban agglomeration in 1970 is the sum of the Chicago standard metropolitan area and the Gary–Hammond–East Chicago standard metropolitan statistical area. Due to U.S. census revisions, urban agglomeration in 1994 is the Chicago–Gary–Kenosha standard consolidated statistical area; thus the population of Kenosha is included, but only in this datum.
[d]Dallas/Fort Worth city proper is the sum of populations of Dallas and Fort Worth in 1973 and Dallas, Fort Worth, and Arlington in 1994. Urban agglomeration in 1973 is the sum of the Dallas and Fort Worth agglomerations. In 1994, urban agglomeration is for the Dallas–Fort Worth–Arlington standard metropolitan statistical area.
[e]Urban agglomeration includes Henderson.
[f]San Francisco Bay area "city proper" is the sum of populations of San Francisco, Berkeley, Oakland, San Jose, Santa Rosa, and Vallejo. Urban agglomeration in 1973 is the sum of the agglomerations of the San Francisco–Oakland, San Jose, and Santa Rosa standard metropolitan statistical areas. Urban agglomeration in 1994 is the San Francisco–Oakland–San Jose standard metropolitan statistical area, comprising Oakland, San Francisco, San Jose, Santa Cruz–Watsonville, Santa Rosa, and Vallejo–Fairfield–Napa.

local census data or direct U.S. census data because we wanted to demonstrate methods that would be applicable around the world and that did not rely on data readily available only for the United States.

3.5 RESULTS AND DISCUSSION

The accuracy in registering the images to one another depended on the degree to which each image was taken looking toward nadir, relative scale, sharpness of focus, degree of contrast of road intersections used as tie points, and other aspects of the images that affected our ability to match locations appearing in each image. The mean of total RMS errors for the image registrations was 3.95 pixels (SD = 3.42, median 2.52, range 1.70 to 10.66). Dallas/Fort Worth was the only city for which we were unable to keep the total RMS error below 5 pixels, and this was because we had to use 9 points with individual RMS errors > 10 and a point with the maximum RMS error of 14.6 pixels. In contrast, for the other images, the worst individual control point RMS errors were 3.11 (Chicago), 3.47 (Mexico City), 4.44 (Las Vegas), 4.52 (San Francisco Bay area), and 9.90 (Vancouver).

Pixel counts, scale factors, and area estimations are shown in Table 3.2, and maps of the identified areas are shown in Figure 3.5; see the color insert). We digitized images taken with both 100- and 250-mm lenses at 1200 dpi and found pixel sizes ranging from 53 to 201 m (Figure 3.6). These pixel sizes are compared to urban area parcel sizes modified from the illustration of Jensen (1983:Figure 30-32) [originally published in a review by Welch (1982:Figure 9)] to compare the area on the ground represented by a pixel from digitized astronaut hand-held photographs with instantaneous fields of view for multispectral scanners (Figure 3.6). Because our objective was to quantify city growth broadly, the pixel sizes in this study do not reflect the best achievable resolution for examination of cities using astronaut photography.

Although built-up area mapped using our methods would differ somewhat from urban areas defined for other purposes (e.g., for the U.S. Census), we wanted to compare our results with those from other sources. The UN estimated 1990 urban areas for city limits and urban agglomerations for all the cities in our study except Mexico City (see details of our compilation of UN data in Table 3.1). We expected that our built-up area estimation would fall between the more restrictive city-limits estimate and the broadly defined urban agglomeration. We also recognized that there could be significant differences between the areas we identified as built-up and the agglomerations defined by the UN. Our estimate fell within the range of UN city and urban agglomeration areas for all cities except the San Francisco Bay area. The area estimated was in the low end of the UN range for Las Vegas and Vancouver, at the center of the range for Dallas/Fort Worth, and at the maximum side of the range for Chicago (Figure 3.7). These results suggest that our method was generally a con-

TABLE 3.2 Images, Built-up Areas, and Pixel Width (Scale) for North American Cities in This Study

City	Baseline Image		Shuttle–Mir Image		% Increase	Pixel Width[b](m)
	Number, Date[a]	Built-up Area [km² (pixels)]	Number, Date[a]	Built-up Area [km² (pixels)]		
Chicago	SL3-46-199, September 18, 1973	1675.0 (170,780)	NM21-767-68, July 1996	2684.4 (273,702)	60.3	99.0
Dallas	AS9-21-3299, March 1969	1418.8 (147,818)	NM22-774-19, February 22, 1997	2872.4 (299,255)	102.4	98.0
Las Vegas	SL3-28-059, August 11, 1973	206.3 (60,917)	NM22-725-34, October 23, 1996	628.3 (185,481)	204.5	58.2
Mexico City	AS9-19-3011, March 1969	574.3 (201,240)	NM22-741-54B, December 1996	1219.5 (427,279)	112.3	53.4
San Francisco	SL3-121-2374, July–September 1973	3994.4 (98,869)	NM21-741-15A, May 5–6, 1996	7125.8 (176,379)	78.4	201.0
Vancouver	SL3-122-2512, July–September 1973	272.1 (24,930)	NM22-703-120, July 23, 1996	385.7 (35,335)	41.7	104.5

[a]Exact date not known for all images. Hand-held cameras used during Apollo 9 (AS9) and Skylab 3 (SL3) did not include technology to imprint exact date and time. This feature malfunctioned for some hand-held shots from Shuttle–*Mir*. Dates are available for photographs from the Skylab fixed multispectral camera (National Aeronautics and Space Administration, 1974) and are included.

[b]Approximate pixel size estimated by measuring the following runways of known length: Chicago CGX 18-36 (3899 ft), Dallas/Fort Worth DFW 13L-31R (9000 ft), Las Vegas LAS 7L-25R (14,512 ft), Mexico City MMX 5R-23L (12,795 ft), San Francisco SFO 10L-28R (11,870 ft), and Vancouver YVR 8L-26R (9940 ft) (Aircraft Owners and Pilots Association, 1997).

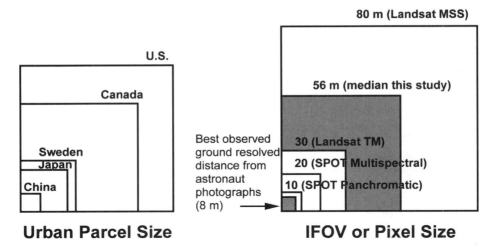

Figure 3.6 Comparison of urban land parcel sizes and IFOVs of satellite remote sensors with pixel sizes for the analysis of digitized photographs (1200 ppi) from low Earth orbit in this study. Urban parcel size represents the estimated size of a parcel of land that would include a single land-use class (Welch, 1982). Smaller pixel sizes than those observed in this study could be obtained by digitizing at higher resolution. For reference, the minimum ground resolved distance observed from the best astronaut photographs is also shown. (Modified from Welch, 1982: Fig. 4; also reproduced in Jensen, 1983: Fig. 30-32.)

servative one. We delimited only the most built-up areas and did not measure lower-density areas. Such low-density areas are linked economically to the city and often counted as part of the urban agglomeration in U.S. Census or UN compilations.

Our estimate of built-up area in the San Francisco Bay area far surpassed the urban agglomeration area reported by the UN because of the sprawling built-up area identified during photointerpretation. [See dynamic visualization from other data sources in Bell et al. (1995); discussion of conversion of land from rangeland to urban use in

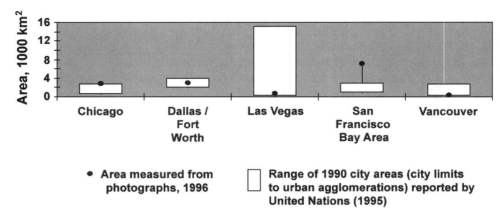

Figure 3.7 Comparison of 1990 urban areas estimated by a special United Nations survey (*Demographic Yearbook*, 1995) with the built-up area measured in this remote-sensing study. The lower edge of the white box indicates the city limits area and the upper edge indicates the urban agglomeration area (details of compilation in Table 3.1).

Forero et al. (1992).] Around San Francisco Bay we observed that cities such as Sonoma, Napa, Vacaville, Fairfield, Petaluma, and Santa Rosa were part of a contiguous built-up area. Most of these communities were included in the population data, but it appears that a less expansive definition of the San Francisco urban agglomeration was used for the 1990 city area data. [The UN *Demographic Yearbook* (1995) provides no details of the composition of the urban agglomerations for area estimates.] Another substantial and contiguous built-up area along the Interstate 680 corridor (Concord and Pleasant Hill to Pleasanton and Livermore) was not included in population data or in city area compilations by the UN (notes to Table 3.1). Although the UN *Demographic Yearbook* (1995) did not compile area estimates for Mexico City, Rapoport et al. (1983) considered the urban area of Mexico city to be 660 km^2 in a periurban area encompassing 988 km^2. Our built-up area estimates of 668 km^2 in 1973 and 1071 km^2 in 1996 (Table 3.2) compare reasonably with these data.

Many of the cities we studied had an increase in built-up area that was disproportionately large relative to their increase in population (Figures 3.8 and 3.9). Notably, the pattern of increase in built-up area followed three patterns. Expansion rates for Dallas, Chicago, Mexico City, and Las Vegas were of the same order of magnitude (53.8, 43.9, 23.9, and 18.3 km^2/yr, respectively). Although it is generally considered part of a region experiencing very rapid rates of growth (e.g., Artibise et al., 1997), Vancouver had a much lower rate of expansion than did the other cities, at 4.9 km^2/yr. Results for the San Francisco Bay area differed significantly from the other cities, with an expansion rate of 136.1 km^2/yr. This result should be considered in light of the expansive

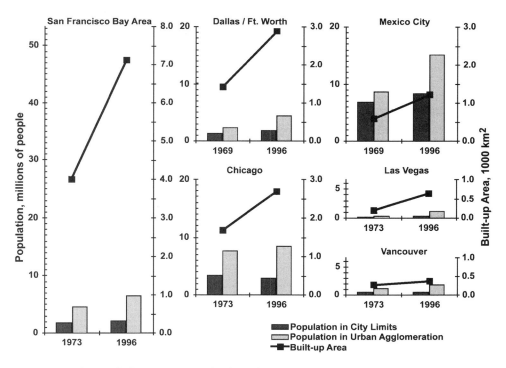

Figure 3.8 Change in built-up area measured in this study compared to increase in population size as tabulated by the United Nations (Table 3.1). To facilitate comparison, data for all cities are presented at a uniform scale for population and area.

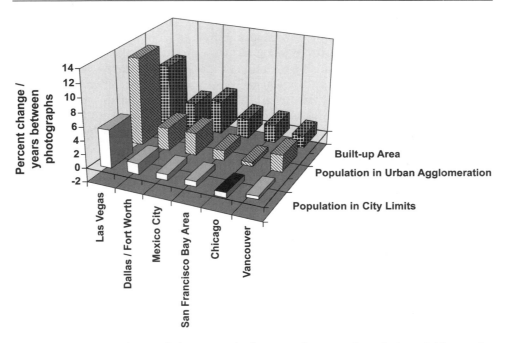

Figure 3.9 Percentage change in built-up area and in human population (population data compiled from UN figures as described in Table 3.1). The total percentage change has been annualized by dividing by the number of years between photographs.

built-up area for San Francisco identified in this study (Figure 3.5). The pattern of growth in built-up area around San Francisco Bay is related partly to topography (constrained by the bay, the Pacific Ocean, and coastal ranges; see also Forero et al., 1992; Kirtland et al.; 1994, Bell et al., 1995) compared to more circular patterns of expansion of the other cities (Figure 3.5). It is also possible that extensions of the city were more dense and therefore more likely to be classified as built-up areas by our methods.

The unique urbanization characteristics of Las Vegas are apparent when the data are viewed as percentages (Figure 3.9). Although its absolute annual growth in area was similar to that of other North American cities (discussed above), Las Vegas had a dramatically higher relative rate of growth (8.9%/yr increase in built-up area compared to an average for the other five cities of 3.2%/yr, SD = 1.17). The built-up area of Las Vegas tripled from 1973 to 1996 while the population within the city limits increased by 127% and the population in the urban agglomeration nearly quadrupled. Thus, its overall amount of growth was comparable to that of other large cities, but this growth was from much smaller initial (1973) values (Figure 3.9).

For all cities other than Las Vegas, the expansion rate of built-up areas exceeded the expansion in the population urban agglomeration (Figure 3.9). This is consistent with a pattern of urban sprawl noted throughout the United States (Stearns and Montag, 1974; Weeks, 1981). The disproportionate increase in built-up area relative to population growth is most notable for Chicago, where the population increased by only 12% in the urban agglomeration (from 1973 to 1996) and decreased by 18% in the city limits, while the built-up area increased by 60%. Las Vegas is probably an exception to this pattern because of the extremely broad urban agglomeration defined by the U.S. Census in

1994 (UN, 1995) which included Henderson and discontinuous unincorporated communities that were not within the contiguous built-up area measured by our methods.

The pattern of urban sprawl for Mexico City was different from the pattern for other cities. Although built-up area did increase disproportionately with population when examined as percentages (Figure 3.9), the actual magnitudes of change reflected the differences between Mexico City and other North American cities in age, density, and other factors (Rocha and Greene, 1997). The substantial growth in population from 1969 to 1996 (Table 3.1) was accompanied by less increase in built-up area than was typical for the other cities in Figure 3.8.

Our calculations of built-up area were not completely correlated with other measures because of the disparity between our spatially oriented approach and the urban area definitions used in UN tabulations (which consider economic linkages in defining urban agglomerations). Because we were relatively familiar with urbanization histories in the areas we studied, we could identify aspects of the underlying census data that contributed to disparities with our calculations of built-up areas.

Another measure of urban population that is used in urban planning is density. For illustrative purposes, we combined our measures of built-up area with UN census figures to estimate population densities and changes in density (Figure 3.10). Our method

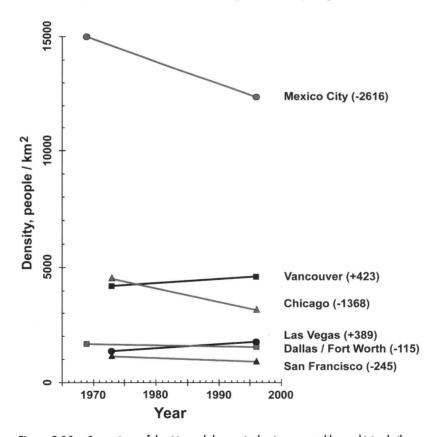

Figure 3.10 Comparisons of densities and changes in density computed by combining built-up area measurements form this study with UN population data (Table 3.1). The number after each city name indicates the magnitude of change over the time period.

provided estimates of changes in density that could be of value for future studies. For example, two cities (Vancouver and Las Vegas) increased slightly in density, while others decreased or stayed relatively constant. Although a detailed statistical analysis of such trends is beyond the scope of this analysis, Figure 3.10 illustrates the potential of this method for providing such measures for subsequent research or analysis.

3.6 CONCLUSION

We present a relatively rapid assessment method for determining built-up area using a time series of photographs taken from low Earth orbit. Our method was designed to use minimal image-processing techniques while allowing measurement of the areal extent of urban growth. The photographs we used are available in the public domain and inexpensive. More important, they are well suited to georectification and to the quantification of changes in urban area. By restricting the external data sources to those readily available for areas worldwide, the method that we developed is applicable to urban growth anywhere in the world, regardless of the availability of maps and demographic data.

Comparison of our measurements of built-up area with reported population sizes and urban areas tabulated by the UN indicates that this method quantifies urban change in a way that could provide answers for certain types of questions. For questions of land-use change relating to urban ecology, conservation biology, and large-scale urban planning, our measure of built-up area could provide an appropriate measure of urban impact on the surrounding environment. The variable change in areal extent with change in urban population could be evaluated in terms of geographic, economic, and social (historical and cultural) factors. Our method for quantifying urban change would be particularly useful in studies that require observations for a large number of cities.

The time series of photographs that astronauts have taken of cities over the last 30 years is a large and relatively untapped database on urban change around the globe. Astronaut photography and remote-sensing techniques can be combined in a new method for study of urban areas worldwide.

ACKNOWLEDGMENTS

We thank D. Pitts and C. Evans for encouragement to conduct this project. Technical assistance was provided by L. Anenson, J. Caruana, C. Cloudt, M. Lozano, and L. Upchurch. D. Amsbury, P. Dickerson, C. Evans, D. Pitts, and R. White provided comments and reviews that improved the manuscript significantly.

REFERENCES

Aircraft Owners and Pilots Association. 1997. *AOPA Airport Directory,* 1998 ed. Frederick, MD: AOPA.
Alig, R. J., and Healy, R. G. 1987. Urban and built-up land area changes in the United States: an empirical investigation of determinants. *Land Economics,* 63:215–226.

Anderson, D. E., and Anderson, P. N. 1973. Population estimates by humans and machines. *Photogrammetric Engineering and Remote Sensing*, 39:147–154.

Anderson, J. R., Hardy, E. E., Roach, J. T., and Witmer, R. E. 1976. *A Land Use and Land Cover Classification System for Use with Remote Sensor Data*. Professional Paper 964. Washington, DC: U.S. Geological Survey.

Artibise, A., Moudon, A. V., and Seltzer, E. 1997. Cascadia: an emerging regional model. In Geddes, R., ed., *Cities in Our Future*. Washington, DC: Island Press, pp. 149–174.

Au, K. N. 1993. *Cities of the World as Seen from Space*. Hong Kong: Geocarto International Centre.

Barnsley, M. J., Sadler, G. J., and Shepherd, J. W. 1989. Integrating remotely sensed images and digital map data in the context of urban planning. *Proceedings, Annual Conference of the Remote Sensing Society*, Vol. 15, pp. 25–32

Bell, C., Acevedo, W., and Buchanan, J. T. 1995. Dynamic mapping of urban regions: growth of the San Francisco/Sacramento region. *Proceedings, Urban and Regional Information Systems Association*, pp. 723–734. Available on-line at: *http://edcwww2.cr.usgs.gov/umap/pubs/urisa_cb.html* (viewed March 26, 1998)

Brown, D. M. 1997 (September 30). *Understanding Urban Interactions: Summary of a Research Workshop*. Arlington, VA: National Science Foundation. Available on-line at: *http://www.nsf.gov/pubs/1998/sbe981/be981.htm* (viewed April 14, 1998)

Campbell, J. B. 1996. *Introduction to Remote Sensing*, 2nd ed. New York: Guilford Press.

Castiaux, N., Massart, M., and Wilmet, J. 1991. Environmental study of tropical African urban areas by multitemporal satellite imageries (Lubumbashi in Zaire, Central Africa). *Proceedings, International Symposium on Remote Sensing of the Environment*, Vol. 24, pp. 161–218.

Christenson, J. W., Dietrich, D. L., Lachowski, H. M., Stauffer, M. L., and McKinney, R. L. 1979. Urbanized area analysis using Landsat data. In *Computer Mapping in Natural Resources and the Environment: Including Applications of Satellite-Derived Data*. Cambridge, MA: Laboratory for Computer Graphics and Spatial Analysis, Harvard University, pp. 5–11.

Cicero-Fernández, P. 1997. Air pollution trends and controls in North America: Los Angeles, Mexico City, and Toronto. In Macari, E. J., and Saunders, F. M., eds., *Environmental Quality, Innovative Technologies, and Sustainable Economic Development: A NAFTA Perspective*. New York: American Society of Civil Engineers, pp. 41–48.

da Costa, S. M. F. 1991. Orbital remote sensing applied to urban environmental impact: a case study. *Proceedings, International Symposium on Remote Sensing of the Environment*, Vol. 24, pp. 209–218.

Earth from space—search (cities). (undated). *http://earth.jsc.nasa.gov/city.html* (viewed October 15, 1998)

Earth Science Branch, NASA Johnson Space Center. 1977 (November 12). Earth Science Branch photographic database. *telnet sseop.jsc.nasa.gov* (username: PHOTOS; viewed February 16, 1998) [*Note*: Support of this search service was discontinued in June 1998. See Office of Earth Sciences (1998) for current alternatives.]

Ehlers, M., Jadkowski, M. A., Howard, R. R., and Brostuen, D. E. 1990. Application of SPOT data for regional growth analysis and local planning. *Photogrammetric Engineering and Remote Sensing,* 56:175–180.

ERDAS, Inc. 1997. *ERDAS Field Guide.* Atlanta, GA: ERDAS.

Forero, L., Huntsinger, L., and Clawson, W. J. 1992. Land use change in three San Francisco Bay Area counties: implications for ranching at the urban fringe. *Journal of Soil and Water Conservation,* 47:475–480.

Friedman, S. Z. 1980. *Mapping Urbanized Area Expansion Through Digital Image Processing of Landsat and Conventional Data.* JPL Publication 79-113. Pasadena, CA: Jet Propulsion Laboratory, National Aeronautics and Space Administration.

Geddes, R., ed. 1997. *Cities in Our Future.* Washington, DC: Island Press.

Hill, J., and Hostert, P. 1996. Monitoring the growth of a Mediterranean metropolis based on the analysis of spectral mixtures: a case study on Athens (Greece). In Parlow, E., ed., *Progress in Environmental Remote Sensing Research and Applications.* Rotterdam, The Netherlands: Balkema, pp. 21–31.

Holz, R. K., Huff, D. L., and Mayfield, R. C. 1969. Urban spatial structure based on remote sensing imagery. *Proceedings, International Symposium on Remote Sensing of the Environment,* Vol. 6, No. 2, pp. 819–830.

Howarth, P. J., and Boasson, E. 1983. Landsat digital enhancements for change detection in urban environments. *Remote Sensing of the Environment,* 13:149–160.

Jensen, J. R. 1979. Spectral and textural features to classify elusive land cover at the urban fringe. *Professional Geographer,* 31:400–409.

Jensen, J. R. 1983. Urban/suburban land use analysis. In Estes, J. E., ed., *Manual of Remote Sensing,* 2nd ed. Vol. II: *Interpretation and Applications.* Falls Church, VA: American Society of Photogrammetry, pp. 1571–1666

Jensen, J. R. 1986. *Introductory Digital Image Processing: A Remote Sensing Perspective.* Englewood Cliffs, NJ: Prentice Hall.

Kam, R.-S. 1995. Integrating GIS and remote sensing techniques for urban land-cover and land-use analysis. *Geocarto International,* 10:39–49.

Kirtland, D., Gaydos, L., Clarke, K., De Cola, L., Acevedo, W., and Bell, C. 1994. An analysis of human-induced land transformations in the San Francisco Bay/Sacramento area. *World Resources Review,* 6:206–217. Available on-line at: *http://geo.arc.nasa.gov/usgs/WRR_paper.html* (viewed March 26, 1998)

Lins, H. F., Jr. 1976. Land-use mapping from Skylab S-190B photography. *Photogrammetric Engineering and Remote Sensing,* 42:301–307.

Martin, L. R. G., Howarth, P. J., and Holder, G. H. 1988. Multispectral classification of land use at the rural–urban fringe using SPOT data. *Canadian Journal of Remote Sensing,* 14:72–79.

McDonnell, M. J., and Pickett, S. T. A. 1990. Ecosystem structure and function along urban–rural gradients: an unexploited opportunity for ecology. *Ecology,* 71:1232–1237.

Meyer, W. B. 1996. *Human Impact on the Earth.* Cambridge: Cambridge University Press.

Mishra, J. K., Aarathi, R., and Joshi, M. D. 1994. Remote sensing quantification and change detection of natural resources over Delhi. *Atmospheric Environment,* 28:3131–3137.

Møller-Jensen, L. 1990. Knowledge-based classification of an urban area using texture and context information in Landsat-TM imagery. *Photogrammetric Engineering and Remote Sensing,* 56:899–904.

Murphy, D. D. 1988. Challenges to biological diversity in urban areas. In Wilson, E. O., ed., *Biodiversity.* Washington, DC: National Academy Press, pp. 71–76.

National Aeronautics and Space Administration. 1974. *Skylab Earth Resources Data Catalog.* JSC 09016. Houston, TX: NASA Johnson Space Center.

National Science Foundation. 1998. *Urban Research Initiative: The Dynamics of Change in Urban Environments.* Initiative Solicitation for FY 1998. NSF 98-98. Arlington, VA: National Science Foundation.

Nordbeck, S. 1965. *The Law of Allometric Growth.* Michigan Inter-University Community of Mathematical Geographers Discussion Paper 7. Ann Arbor, MI: Department of Geography, University of Michigan.

Office of Earth Sciences, NASA Johnson Space Center. 1998 (July 24). *http://eol.jsc.nasa.gov* (viewed July 30, 1998)

Ogrosky, C. E. 1975. Population estimates from satellite imagery. *Photogrammetric Engineering and Remote Sensing,* 41:707–712.

Paulsson, B. 1992. *Urban Applications of Satellite Remote Sensing and GIS Analysis.* Washington, DC: The World Bank.

Rapoport, E. H., Díaz-Betancourt, M. E., and López-Moreno, I. R. 1983. *Aspectos de la Ecología Urbana en la Ciudad de México.* Mexico City: Editorial Limusa.

Rocha, X. C., and Greene, F. 1997. Mexico City: the city of palaces. In Geddes, R., ed., *Cities in Our Future.* Washington, DC: Island Press, pp. 179–188.

Sadler, G. J., and Barnsley, M. J. 1990. Use of population density data to improve classification accuracies in remotely-sensed images of urban areas. *Proceedings, European Conference on Geographic Information Systems,* 1:968–977.

Spengler, J., and Ford, T. 1997. From the environmentally challenged city to the ecological city. In Geddes, R., ed., *Cities in Our Future.* Washington, DC: Island Press, pp. 33–62.

Stearns, F., and Montag, T. 1974. *The Urban Ecosystem: A Holistic Approach.* Stroudsburg, PA: Dowden, Hutchinson & Ross.

Stow, D. A., Collins, D., and McKinsey, D. 1990. Land use change detection based on multi-date imagery from different satellite sensor systems. *Geocarto International,* 3:3–12.

Toll, D. L. 1985. Effect of Landsat Thematic Mapper sensor parameters of land cover classification. *Remote Sensing of the Environment,* 17:129–140.

United Nations. 1991. *World Urbanization Prospects, 1990.* ST/ESA/SER.A/121. New York: Department of International Economic and Social Affairs, United Nations.

Weeks, J. R. 1981. *Population: An Introduction to Concepts and Issues.* Belmont, CA: Wadsworth.

Welch, R. 1980. Monitoring urban population and energy utilization patterns from satellite data. *Remote Sensing of the Environment,* 9:1–9.

Welch, R. 1982. Spatial resolution requirements for urban studies. *International Journal of Remote Sensing,* 3:139–146.

Wellar, B. S. 1969. The role of space photography in urban and transportation data series. *Proceedings, International Symposium on Remote Sensing of the Environment,* Vol. 6, No. 2, pp. 831–854.

Chapter

4

Fluctuating Water Levels as Indicators of Global Change: Examples from Around the World

Cynthia A. Evans, Joe Caruana, David L. Amsbury, and Kamlesh P. Lulla

Office of Earth Sciences
NASA Johnson Space Center
Houston, Texas USA

ABSTRACT

Astronauts participating in the Shuttle–*Mir* program from 1996 through 1998 documented the interaction between short-term climate variability and longer-term human demands on water resources in many regions of the world. Several high-use watersheds experienced rapid changes and extremes—both floods and drought—during this period. In Asia, astronauts documented the continued desiccation of the Aral′ Sea and effects of flooding along the shores of the Caspian (see Chapters 9 to 19). Southwestern North America went from drought in 1996 to flooding in 1997 and 1998, while the Ohio–Mississippi Valley flooded in spring of 1997. In Africa, heavy monsoons flooded Lake Nasser in late 1996 to its highest level ever, resulting in new developments in the Egyptian desert. Similarly, heavier-than-usual precipitation in early 1998 filled new reservoirs on the Euphrates River in Turkey, and El Niño–related rains in early 1998 flooded much of northeast Africa. We also tracked water levels in Lake Chad, seasonal changes along the inland delta of the Niger River, and the Okavango swamp in Botswana. Because astronauts on long-duration flights develop detailed knowledge about the status of water bodies, they are able to recognize and record changes. The image data collected during the Shuttle–*Mir* long-duration flights allowed differentiation of seasonal variability, annual climate variations, and some human-induced changes.

Comparative analyses of these images with historical imagery allowed changes to be quantified in several large reservoirs and catchment basins. For example, the Aral′ Sea continued to shrink by roughly 5000 km^2 between 1996 and 1998, now covering an area of only 27,000 km². The area of Lake Amistad on the Rio Grande was reduced by more than half during the 1996 drought in western North America. In Egypt, a newly flooded depression covers 256 km². These images and the simple analyses resulting from them can become tools for rapid assessment of changing conditions in high-use watersheds around the world.

4.1 INTRODUCTION

Water rights and water uses are among today's most important resource issues, particularly as human populations and the consequent demands for water swell in arid and semiarid regions. Global water use has quadrupled in the past 50 years, and today's human population uses about 44% of the available global runoff (Liniger, 1995). International watersheds (watersheds crossing international boundaries) comprise almost 50% of the global land area (Gleick, 1993; Liniger, 1995; Frederick, 1996). Virtually every river or lake in the world that crosses or forms a political border is a focus for controversy surrounding the use and allocation of the water. In some cases, these multinational resources are a potential source of conflict. Some of the most optimistic predictions using growth projections by the United Nations indicate that by 2050, 2 billion people will live under water-stressed conditions (Gardner-Outlaw and Engelman, 1997). As populations increase in regions with limited water supplies (the Middle East or northern Africa, for example), small changes in supply caused by climate variability or land-use changes will have large impacts on the population. Recent well-publicized extremes in weather around the world (e.g., El Niño) and the catastrophic effects of droughts and floods (such as the 1998 summer floods in China and the 1997–1998 droughts in southeast Asia and Brazil) have made us acutely aware that short-term and predicted long-term variability in climate and weather will continue to increase the stresses on the world's freshwater resources during the twenty-first century.

Photographs from the Shuttle–*Mir* program document water-level changes around the globe as well as land-use differences that affect available freshwater supplies. Through our Earth Sciences operations on *Mir* and our review of the continuous astronaut–cosmonaut observations of Earth's surface over the last two years (March 1996 through May 1998), we have been struck by the rapid rates of regional climate change resulting in floods, drought, or both in the same geographic region (Figure 4.1) (Climate Prediction Center, 1998; Dartmouth Flood Observatory, 1998).

The most visible impacts of water use and hydrological modification by human populations from space are irrigated cropland (which uses up to 70 to 80% of the freshwater supplies in many arid regions), human-made lakes, and rapidly variable water levels in certain rivers, lakes, and reservoirs (Fels and Keller, 1973; Gleick, 1993; Liniger, 1995). Most of the world's largest reservoirs and many of the extensive irrigated fields have been created in the past 50 years, generally after World War II (Fels and Keller, 1973; Liniger, 1995).

Land–water boundaries are readily observed by astronauts and cosmonauts orbiting Earth, so the Shuttle–*Mir* photographs contain a wealth of data on the world's rivers, lakes, reservoirs, and ice fields. For the same reason, a robust historical database of astronaut photographs, some as early as 1965, exists for many important water resources. As a time series, the photographs document rapid rates of change

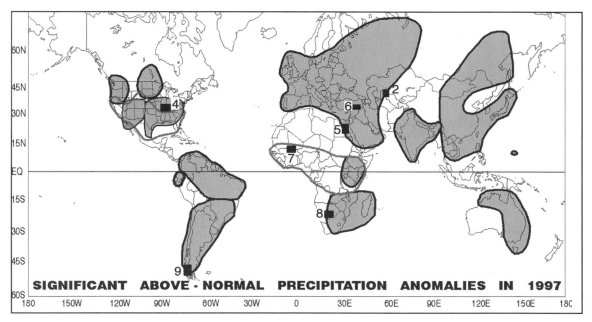
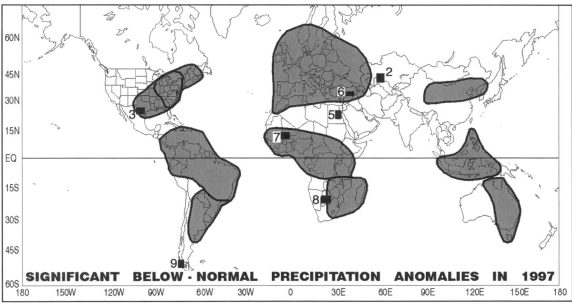

Figure 4.1 Map of significant precipitation anomalies in 1997. The boxes numbered 2 to 9 correspond to Figures 4.2 to 4.9, and identify locations of examples discussed in the text. (Modified from Climate Prediction Center, 1998.)

related to short-term climate variability and changes in regional land-use and water demands, including the creation and growth of human-made lakes. Perhaps most importantly, the photos provide the larger geographic context for rivers, lakes, reservoirs, and their watersheds and a means of visualizing the regional changes related to water use and water abundance.

In this chapter we use Shuttle–*Mir* photographs to profile several important international water resources which have experienced considerable change over the two years spanning the Shuttle–*Mir* program (1996–1998). Examples include the Aral' Sea in central Asia, Lake Amistad on the Texas–Mexico border, the Ohio–Mississippi rivers, Lake Nasser on the Nile River, the inland delta of the Niger in Mali, and the Okavango swamp in Botswana. Chapters 9 to 19 cover the Caspian Sea. Even though this time period included the extreme 1997–1998 El Niño event (Chapter 5), many of the examples presented here were not linked directly to El Niño effects.

We also show the breadth of water resource topics that can be addressed with the imagery from the Shuttle–*Mir* program and the larger NASA–Johnson Space Center Earth Sciences database (Office of Earth Sciences Database, 1999). Another purpose of this paper is to demonstrate how simple desktop image-analysis methods using software that is inexpensive and commonly available can be applied to the image data to obtain first-order quantitative results. These measurements can be used for rapid assessment of changing conditions of many of the world's watersheds.

4.2 METHODS

A variety of image-analysis techniques were used to obtain quick and quantitative measurements of change from the photographs shown in this paper. All comparative analyses used at least two images of the same region to comprise a time series. In some examples, the Office of Earth Sciences database (1999) was queried for an appropriate baseline image to compare with an image from the Shuttle–*Mir* program. All photographs were electronically scanned from the film using an Agfa Arcus II scanner at the minimum resolution necessary for analyses (generally, 600 pixels per inch for a 55 × 55 mm film positive). The best maps available of the same regions were also scanned—the maps came from a variety of sources.

Image sets and maps were electronically cropped and resized to a common area, file size, and resolution using Adobe Photoshop. The images were individually rectified to the respective map base in Photoshop by co-registering landmarks across the images. In all cases the two images were also co-registered to each other using common landmarks not marked on maps. This empirical method of registration was quick and interactive and provided for more detailed manipulation of certain types of changing features (e.g., vegetation in river distributaries) that were not well defined on the maps.

Depending on image clarity and contrast, the rectified images were contrast-enhanced in Photoshop and then exported into NIH Image (NIH Image home page, 1998). In NIH Image, additional image enhancement was performed, and the region of interest and pixel scale were defined allowing for areal calculations. Each measurement was made three to five times, and the median value was used. Differences in measurements were due to variability in scene contrast and definition of boundaries. Total differences in areal measurements were less than 10% for small areas (e.g., Lake Amistad) and 1% for large areas (e.g., Aral' Sea). Our calculations were consistent with other estimates, when available (e.g., Micklin, 1988, 1991).

4.3 REGIONAL EXAMPLES

4.3.1 Aral' Sea

One of the most publicized examples of human-induced environmental changes related to an overuse of freshwater resources is the Aral' Sea in central Asia. Even the current Vice President of the United States has written about the Aral' Sea tragedy (Gore, 1992). The Amu Darya and Syr Darya (the rivers running into the Aral') are the only freshwater resources in the region. These rivers include international borders between six different countries and have become emblems of the complications arising from excessive regional water demands and international water issues.

The Aral' Sea lies in an arid basin in central Asia within the boundaries of the former Soviet Union. Beginning in the 1950s, ambitious programs to harness the regional water resources resulted in the extensive irrigation and development of cultivated lands along the two major rivers feeding the Aral' Sea (Micklin, 1988, 1991; Glantz, 1994). Because of the increasing diversion of water to support regional agriculture, the Aral' Sea level began to drop dramatically. In the late 1980s, the sea had divided into two separate basins as the result of dropping water levels. By 1990, the surface area of the Aral' Sea had decreased by more than 50%, and the salinity had tripled since 1960, when the sea covered about 68,000 km^2 (Micklin, 1988, 1991, 1993; Glazovskiy, 1995; Middlet, 1996). In addition to the simple reduction of water area in the Aral' Sea, irrigation and agricultural practices that affect water quality have resulted in several other changes in and around the Aral' Sea. The fish and plankton populations in the sea have died, causing the demise of the fishing industry, and the Amu Darya and Syr Darya deltas have suffered drastic changes in their ecology, including desertification. Other impacts on the regional land use include salinization of the soils due to rising groundwater levels in irrigated regions, regional climate change, and increased human health problems related to declining environmental quality (Glazovskiy, 1995).

Photographs taken by Shuttle–*Mir* crew members in mid-1996 and early 1998 indicate that the water level of the Aral' Sea continues to drop quickly, even though general precipitation levels were normal or above for the region in the second half of 1997 (Climate Prediction Center, 1998) (Figure 4.1). Figure 4.2 (see the color insert) shows the relative water levels in the Aral' Sea in 1985, 1992, 1996 (at the beginning of the Shuttle–*Mir* program), and early 1998. The photographs in Figure 4.2 show that Barsakel'mes Island (d) near the middle of the sea has joined with the eastern shoreline, and the southern basin is now nearly bisected into two smaller basins by the growth of Vozrozhdeniya Island (c). Figure 4.3 shows the outline (mapped to a common map base) of the Aral' Sea over time taken from the photographs in Figure 4.2 (1985 and 1992 rectification, K. Knowlton, unpublished data). The calculated area of the Aral' Sea has decreased from 48,500 km^2 in 1985 to roughly 27,000 km^2 in January 1998. Our calculations for 1985 and 1992 areas agree with other published estimates of the Aral' Sea area (e.g., Micklin, 1988; 1991). The 1996 and 1998 areal calculations are new data obtained from the Shuttle–*Mir* photographs.

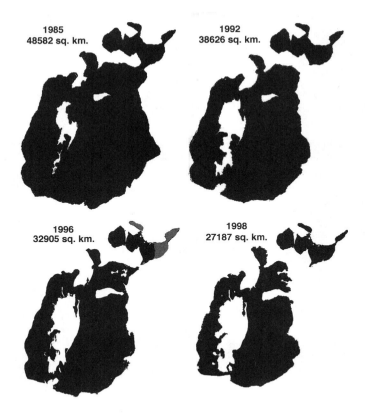

Figure 4.3 Areal coverage of the Aral' Sea, mapped from the photographs in Figure 4.2. The images were rectified to a common map base (K. Knowlton, unpublished data), and areas were measured from these projections.

4.3.2 Lake Amistad, Texas–Mexico Border

The border between the United States and Mexico runs through arid country for more than 3000 km; the Rio Grande forms more than 2000 km of that border. The two countries have multiple treaties regarding water use of the Rio Grande, and the International Boundary Water Commission (IBWC) administers the use of the boundary waters. A relatively recent joint venture was the 1969 construction of the Amistad Dam at the confluence of the Pecos and Devils Rivers with the Rio Grande near Del Rio, Texas. Lake Amistad, the reservoir behind the dam, became the regional water resource as well as a National Recreation Area (International Boundary Water Commission, 1998; National Park Service, 1998).

The southwestern United States and northern Mexico experienced a severe drought from October 1995 through August 1996. During the summer of 1996, Shuttle–*Mir* crew members documented extremely low water in reservoirs that were affected by the drought, such as the Amistad Reservoir (Figure 4.4) and the Obregon Reservoir on the Yaqui River in Sonora, Mexico. Much of the region received only

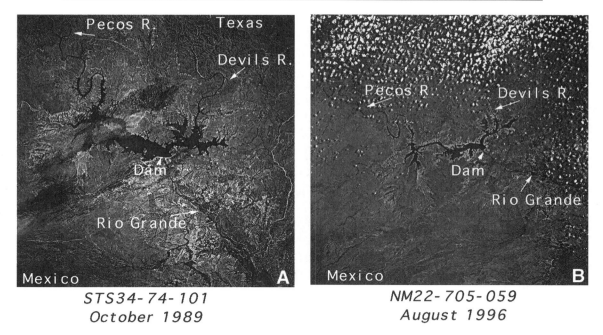

Figure 4.4 Amistad Reservoir, Texas–Mexico border. The 1996 drought in the southwestern United States resulted in a 50-ft drop in water level of Amistad Reservoir. View (*B*), taken in August 1996, shows the former shoreline of the lake (white region circling the reservoir) and its diminished size (calculated to cover roughly 75 km²) after the drought (NASA photograph NM22-705-059). For comparison, view (*A*), taken in October 1989, shows normal water levels covering an area of nearly 200 km² (NASA photograph STS034-74-101).

25 to 50% of the normal rainfall (National Oceanic and Atmospheric Administration, 1996a, b). By June 1996, water levels in the reservoirs along the Rio Grande in Texas and Mexico had dropped to record lows. Figure 4.4 shows Lake Amistad at its lowstand, 50 ft below normal. For comparison, we include an image from 1989, when water levels were normal. We registered the images to a common map base, and calculated the approximate area of the reservoir in each image. These calculations indicate that by the end of the drought in August 1996, the surface area of Lake Amistad was about 75 km², down from roughly 200 km² in 1989.

4.3.3 Ohio–Mississippi River Floods, March 1997

Extensive flooding of the Ohio and Mississippi Rivers occurred in early March 1997 and was well documented by the NASA 4 crew (Figure 4.5; see the color insert). The Ohio River started cresting and flooding regions in West Virginia, Kentucky, Ohio, and Indiana on March 3, 1997. Floodwaters reached their highest levels since 1964 and inundated the cities of Louisville, Kentucky and Cincinnati, Ohio. Across the region, tens of thousands of people were evacuated, and the river was closed to commercial traffic (CNN News, 1997; CNN Weather, 1997a–d). Shuttle–*Mir* crew members started recording movement of the floodwaters cresting around Evansville,

Indiana, on March 9, 1997 (NM23-705-320 and 321, Figure 4.5). Subsequent photographs traced the floodwaters to the confluence of the Mississippi and Ohio (NM23-712-434 to 438, Figure 4.5) on March 15. By that time, the crest was reaching Memphis, Tennessee. Again on March 18, the crew photographed the vicinity of St. Louis (NM23-713-310 to 314, not shown), and continued to document the detail the flooding at the Ohio–Mississippi river junction on March 23 (NM23-728-420-429, not shown). Although the severity of flooding diminished in the much wider Mississippi River, the pulse of high water was traceable in the lower Mississippi almost to the Gulf of Mexico on March 21 (NM23-716-368, not shown).

The floods were also well documented using imagery from NOAA satellites, and remote sensing is now routinely used to study such floods (Brakenridge et al., 1994). However, the NOAA satellites cannot achieve the resolution of the astronaut/cosmonaut photography. For example, NM23-712-438 (Memphis image, Figure 4.5) was taken obliquely with a 100-mm lens from an altitude of 380 km and has a calculated resolution of approximately 115 m. Also, the time series of images taken by the astronauts are in natural color and from different viewing angles and allow viewers to appreciate how the floodwaters move downstream in a pulse. Images of the floods and the downstream runoff and sediment plume into the Gulf of Mexico have the potential to contribute to a broad range of studies regarding the hydrologic responses to heavy rainfall and snowmelt, changes in land-use practices, and the downstream impacts of significant runoff. For example, Mississippi River floods have been correlated to downstream events such as the growth of the hypoxic zone ("dead zone") and red tides in the Gulf of Mexico (Goolsby and Battaglin, 1995; Showstack, 1997). The Ohio–Mississippi flood sequence stands out as an example of how the movement of floodwater was tracked for several days. Other large floods have been recorded during the Shuttle–*Mir* program, including floods in the Paraná River drainage in Argentina and Uruguay in April 1997 and again in early 1998 (see Figure 5.2).

4.3.4 Lake Nasser, Egypt

In contrast to floods in the Ohio–Mississippi Rivers and overuse of water in the Aral' Sea region, floods in the Nile River and Lake Nasser in September 1996 contributed to a closer examination of Egypt's use of its allocation of Nile River water and the resurrection of different plans to transport Nile River water into the Egyptian desert (Figure 4.6). Lake Nasser, one of the world's largest reservoirs, was built to ensure Egypt's water conservation and storage and to provide flood protection. The lake began filling in the 1960s after the construction of the Aswan High Dam in central Egypt (Fels and Keller, 1973; Raheja, 1973).

Most of Egypt's 60 million people live in the roughly 5% of arable land composed of the Nile valley and delta (Myllylä, 1995; CNN World, 1997c), downstream from Lake Nasser. Although Egypt negotiated its portion of Nile River water through an international treaty with Sudan in 1959 (McCaffrey, 1993), Egypt's rapid rate of growth and development demand new strategies for supporting the population into the twenty-first century. Recently, the Egyptian government implemented plans for more efficient use of the water stored in Lake Nasser. By diverting water from Lake Nasser, Egypt hopes to claim up to 25% of the nation's land for settlement and agriculture (Egypt Online State Information Service, 1997b).

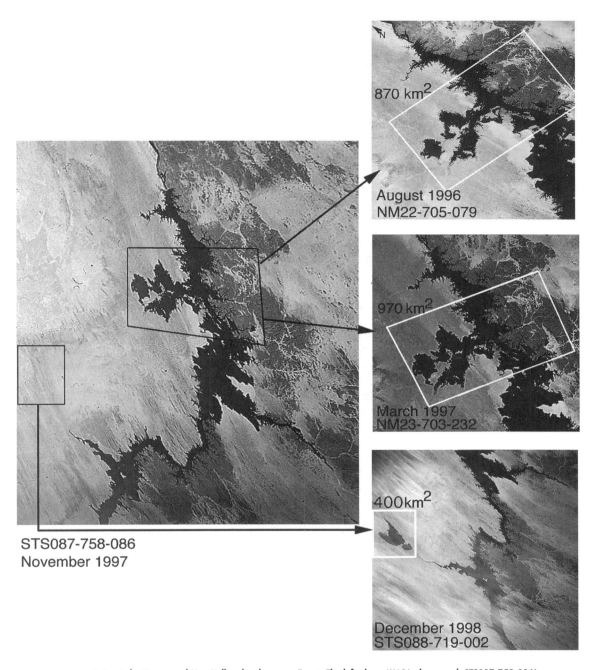

Figure 4.6 Lake Nasser and New Valley development, Egypt. The left photo (NASA photograph STS087-758-086) is an oblique view showing most of Lake Nasser in November 1997. On the right, the top two photos (NASA photograph NM22-705-079 and NM23-703-232) compare the relative water levels in the same section of central Lake Nasser in August 1996 and March 1997. The white boxes define the section of Lake Nasser analyzed in this paper. The lake areas within the box are 870 km² in August 1996 and 970 km² in March 1997. After the rise in lake level, the New Valley Project (bottom right, NASA photograph STS088-719-002) was implemented, tapping Lake Nasser from the west and transporting water into a formerly barren wadi. The project was not visible as late as June 7, 1998. The December 1998 photograph shows the recently flooded New Valley. The area of the flooded region is calculated to be about 400 km².

Figure 4.6 illustrates how astronaut photography can be used to monitor changes in regional hydrology and water distribution. Heavy monsoon rains in the Ethiopian source of the Nile in late August and early September 1996 flooded the Nile, and Lake Nasser rose to its highest levels ever. Precipitation in 1997 and 1998 maintained the highstand in Lake Nasser. The high water level prompted Egyptian officials to examine ways to utilize more fully the Nile water stored in Lake Nasser (CNN World, 1997a; Werner, 1997). Several projects, many of them in various stages of planning since the 1960s, have been announced. Two of the largest programs will irrigate hundreds of thousands of acres in a wadi northwest from Lake Nasser (CNN World, 1997c; Egypt Online State Information Service, 1997a) and in the central Sinai Peninsula (CNN World, 1997a, c). Other projects include the development of an abandoned river tributary between Cairo and Alexandria west of the current delta (Egypt Online State Information Service, 1998). These projects are controversial. Egypt's upstream neighbors have concerns about the total water budget and water oversubscription, and the regional environmental impacts of the large-scale development (McCaffrey, 1993).

Reservoirs and irrigated land are easy to observe and document, especially in arid regions. The astronaut–cosmonaut photographs provide a means of rapid assessment of ongoing water diversions in places such as the Egyptian desert. The Shuttle–*Mir* and Shuttle crews were able to photograph the Nile in Egypt and Sudan repeatedly from 1996 to 1998 (Figure 4.6). During NASA 2 and 3, the astronauts documented the flooding of Lake Nasser in early September 1996. The diversion of water to the New Valley project was photographed later, in 1998. Figure 4.6 shows the unrectified photographs of the water-level changes in and around Lake Nasser. We also rectified each image to an ONC chart and to each other, and calculated the areal coverage of the floodwater along one portion of Lake Nasser (marked on Figure 4.6). There was an increase of approximately 100 km^2 along this section of Nasser between August 1996 and March 1997. We also calculated that more than 400 km^2 of dry land has recently been flooded in the New Valley (bottom image, Figure 4.6).

Other photographs of the region taken from *Mir* cover the lower Nile from north of Khartoum to the Nile delta. The lower Nile has been photographed hundreds of times since the early human spaceflight, so the land-use and hydrological changes since 1965 are well chronicled. The Shuttle–*Mir* images of the delta region (not shown here) document intense agricultural and industrial activity in the delta and continued growth and development west of the delta, including the production of smoke and smog palls (see Chapter 6). These photographs now form a baseline for observing future changes due to the proposed waterworks in Egypt and the upstream countries. Astronauts on future Shuttle and International Space Station missions will continue to photograph this region, allowing for continued quantitative assessment of the regional development around the Nile River.

4.3.5 Greater Anatolia Project, Turkey

Another area of political conflict over water rights and water diversion is the Greater Anatolia Project along the Tigris–Euphrates Rivers in Turkey (Figure 4.7). The sources of both the Tigris and Euphrates Rivers are in Turkey. The rivers flow through or along Syria before entering Iraq. Syria is heavily dependent, and Iraq is completely dependent on the water from these two rivers. Figure 4.7 shows a regional view of an extensive water project on the Euphrates River that will provide irrigation water and electricity to southeastern Turkey. The overall project consists of 13 major initiatives, with 22 dams and 19 hydroelectric power plants. The design will provide irrigation for 1.7 million hectares of land and 27 billion kilowatthours of electricity. The main engineering effort is the Ataturk Dam, now one of the world's largest dams, completed in 1990 (Postel, 1992; McCaffrey, 1993; Turkish Ministry of Foreign Affairs, 1998).

When the entire project is completed in 2005, Turkey will be capable of controlling the flow of the Tigris and Euphrates Rivers (McCaffrey, 1993; Gardner-Outlaw

Figure 4.7 Euphrates River and the Ataturk Dam: (*A*) map showing the upper Euphrates River and location of the photo (box); (*B*) photograph (NASA photograph NASA7-703-56K) showing the recent hydrologic modifications in Turkey and the proximity of the Turkish Ataturk Dam to the Syrian Euphrates Dam (marked "Dam").

and Engelman, 1997). Turkey plans to divert approximately 28% of the river flow; Syria could lose up to 40% of its water, and Iraq could lose up to 90% (McCaffrey, 1993). Future imagery of this region will allow continued monitoring of the ongoing changes in hydrology, including reservoir levels and new irrigation schemes in Turkey, Syria, and Iraq.

4.3.6 Inland Delta of the Niger

Shuttle–*Mir* crew members photographed the Sahara and African Sahel repeatedly, collecting data on seasonal variability of water resources. This region, including the inland delta of the Niger River, has served as a testbed for remote-sensing studies of desertification, land cover, and land use since 1969, when Nimbus III collected imagery of several sites affected by the Sahelian drought prior to the Landsat 1 launch (MacLeod et al., 1977). The Sahelian drought was studied subsequently using Landsat 1 data and then from Skylab 4 in early 1973 (MacLeod et al., 1977). These early studies established a long-term monitoring site for relating regional climate variability and land-use practices. As desertification from both climate variability and intensive land use has proceeded into the 1990s, Earth scientists have been monitoring seasonal variations in water and vegetation abundance in these regions and have documented longer-term trends and fluctuations (MacLeod et al., 1977).

Remote-sensing analysis demonstrates the seasonal differences in water resources and vegetation cover and the strong visual signature of vegetation growth in the swamps of the inland delta of the Niger (Figure 4.8; see the color insert). The inland delta is a large (more than 300 km long) region extending from south-central Mali to the northeast to Timbuktu, where the Niger River flows into a basin of very low relief and spreads out into pools and distributaries. The areal extent of the ephemeral seasonal growth was recorded in late 1996 and early 1997 from *Mir*. The lush vegetation cover lasts only a few months, near the end of the year after summer rains migrate downstream and flood the region. Knowledge about freshwater resources and land cover are important because this part of the Niger also supports one of the largest human populations in the Sahel.

4.3.7 Okavango Swamp, Botswana

The Okavango River and swamp are in another region where controversial water diversions are proposed. The Okavango River is the third largest in Africa—it flows from Angola, forms part of the Angola–Namibia border, and then flows into Botswana, where it spreads into a large inland delta known as the Okavango swamp. Because of its large size (spanning more than 200 km), the regional systematics of the Okavango swamp are best studied from space. The Okavango swamp is easily visible from low Earth orbit, as the vegetation supported by the river is dark and shows up in strong contrast to the surrounding Kalahari desert (Figure 4.9).

The water in the Okavango is a valuable resource to the arid countries of Angola, Namibia, and Botswana. This region suffered drought for much of the last decade, and demands for water resulted in increased regional tension by late 1996–early 1997. The three Kalahari countries had agreed to evaluate the impacts of water supplies and diversions over a 10-year period, but the current drought and Namibia's increasing water needs threaten the agreement. Depleted groundwater and reservoir supplies and a growing population have led Namibia to propose diverting water from the Okavango River at the Angolan border through a pipeline and open canal to the central part of the country (Weekly Mail and Guardian, 1996a, b; CNN World, 1997b; Gardner-Outlaw and Engelman, 1997). The proposed diversion of water into Namibia would threaten the lush Okavango swamp, a wetland wildlife refuge with a delicate ecosystem and Botswana's prime tourist destination.

Shuttle–*Mir* astronauts photographed the Okavango swamp throughout 1996 and 1997, and their images allowed for changes in the regional vegetation coverage in the main part of the delta to be measured. Figure 4.9 features a photograph of the delta taken in October 1996 showing that several of the delta tributaries, usually dark with vegetation, were dry. Rains in early 1997 temporarily relieved the problem, and the vegetation growth in the main part of the delta had increased by late July 1997 from 1840 km^2 (October 1996) to 2750 km^2 (July 1997). Nevertheless, the regional demands on limited water remain, and the long-term drought continued throughout 1997—the region received only 25 to 50% of the normal rainfall

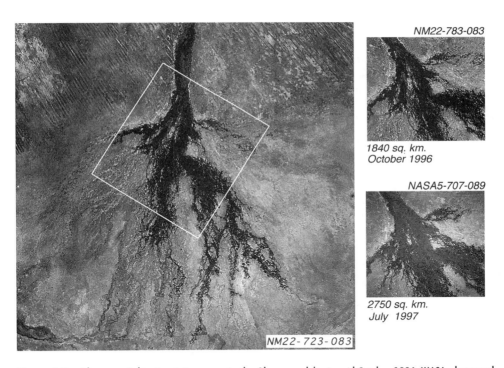

Figure 4.9 Okavango Delta. Vegetation cover in the Okavango delta in mid-October 1996 (NASA photograph NM22-723-083) and early July 1997 (NASA photograph NASA5-707-089). The left-hand view is a synoptic photograph of the Okavango, taken with the 100-mm lens in mid-October 1996. Note that the distal ends of many of the delta fingers are dry, especially on the western side of the river. The white box on the photo depicts the area analyzed in the detailed sections on the right for change in vegetation cover between October 1996 (1840 km^2 of vegetation) and July 1997 (2750 km^2 of vegetation). Although many distributaries are still dry in July 1997, more vegetation filled the center part of the delta and extended farther out the distributaries, compared with the October 1996 view.

amounts because of the strong El Niño event (Climate Prediction Center, 1998). Because of continuing water pressures in the region, southern Africa and the Okavango will remain an important target for future monitoring by Space Shuttle and International Space Station crews.

4.3.8 Ice Fields, Patagonia

The changes in the amount of water retained in snow and ice, and the amount and timing of snowmelt are important for both regional and global water budgets. Changes in these factors are especially important where snowmelt supplies water to arid regions. About two-thirds of the world's fresh water is stored in the form of ice on mountains or in the polar ice caps (Gleick, 1993; Liniger, 1995). A recent focus on the sizes and changes in the world's glaciers and snowfields is related to indications of changing climate (L'vovich, 1979; Liniger, 1995).

Shuttle–*Mir* astronauts had many opportunities to photograph striking views of snow and ice levels and glaciers in regions such as Patagonia, the Alps, the Himalaya, and the Karakoram mountains. Figure 4.10 shows details of several Patagonian glaciers. Although we have not provided a comparative image of the region, the photographs show the exact position of the endpoints of glaciers, clusters of icebergs at the ends of some glaciers, and in the sunglint, differently textured regions of other glaciers, which we interpret as glacier toes covered with meltwater. The photographs serve as examples of how the images taken by astronauts and cosmonauts are data that can be used to quickly assess dynamic conditions in regions where glaciers and snowpack comprise important water resources.

4.4 CONCLUSION

The Shuttle–*Mir* photographs provide a 1996–1998 baseline recording impressive changes in water resources over several regions of the world where freshwater availability and quality are important issues. As a time series, the Shuttle–*Mir* imagery provides a means of rapidly assessing the amount and rates of change experienced by these regions over a two-year period. These data frequently provide more detailed and current information about the changing boundaries of water resources (rivers, reservoirs, ice fields) than do existing maps of these regions. Regional oblique photographs allow for quick assessments of changing water boundaries and vegetation differences. Higher-resolution views (using a 250-mm lens) of many regions provide details over a smaller area for specific change analyses. Sunglint scenes highlight water–land boundaries, which are sometimes very difficult to define. The methods and analyses used here can readily be applied to similar time series of images from the photography database (Office of Earth Sciences Database, 1999) and from imagery acquired from future Shuttle and International Space Station missions to document changes within major watersheds around the world.

During the two years of data collection, several regions around the world experienced significant excesses or deficits from the normal seasonal precipitation. The visible results in changes in water levels, especially in regions where the hydrology has been heavily modified for flood control, water supply and diversion, and irrigation, can be linked with news reports of the resulting human and ecological tragedies.

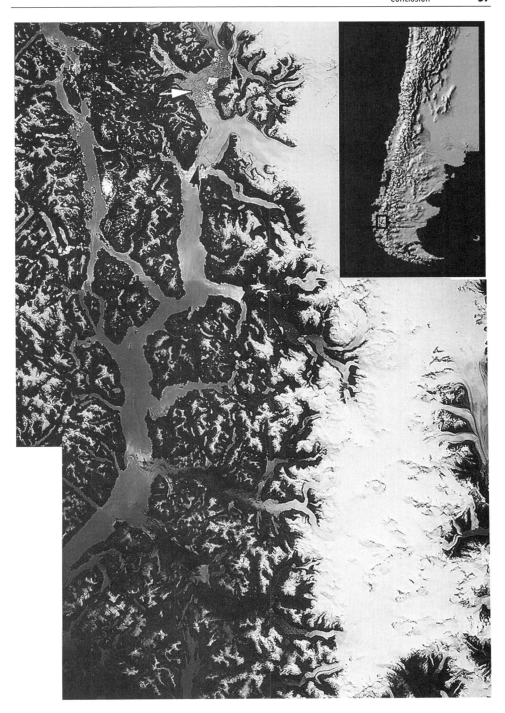

Figure 4.10 Patagonian glaciers. This composite of two photographs (NASA photographs NM22-737-091 and NM22-737-092) taken in the sunglint, shows the position of several glaciers flowing from the ice field of Patagonia (the Darwin Cordillera) into the Pacific fjords. Taken during southern hemisphere summer, the glaciers are calving and the sunglint pattern suggests that there is water covering parts of some of the glaciers (arrows).

Most of the world's important water resources are international. The data presented here indicate that the combined impact of short-term and seasonal climate variability and increasing demands for water (through population increases and modifications of regional hydrology) play a significant role in determining water availability for a given region. These data also provide examples of possible impacts on many key freshwater resources from a longer-term climate change. Thus, the global inventory provided by the Shuttle–*Mir* long-duration missions will be important for monitoring future changes in global freshwater supplies in the twenty-first century.

ACKNOWLEDGMENTS

We are grateful for the help and support for this work from Julie Robinson and Pat Dickerson. We also thank Mike Helfert for his encouragement and thoughtful review—his comments greatly improved the text—and acknowledge K. Knowlton for use of unpublished Aral' Sea data. Figure preparation was facilitated by Leslie Upchurch and Marco Lozano.

REFERENCES

Brakenridge, G. R., Knox, J. C., Paylor, E. D., II, and Magilligan, F. J. 1994. Radar remote sensing aids study of the great flood of 1993. *Eos, Transactions of the American Geophysical Union*, 75:521–527.

Climate Prediction Center. 1998. CPC: analysis and monitoring products. National Oceanic and Atmospheric Administration. *http://nic.fb4.noaa.gov:80/products/analysis_monitoring* (viewed November 11, 1998)

CNN News. 1997 (March 6). Cincinnati still flooded as Ohio surge moves to Louisville. *http://cnn.com/US/9703/06/flood/index.html* (viewed November 11, 1998)

CNN Weather. 1997a (March 4). Waters rising across Ohio Valley. *http://cnn.com/WEATHER/9703/04/ohio.floods.update* (viewed November 11, 1998)

CNN Weather. 1997b (March 5). Muddy Ohio keeps on rising. *http://cnn.com/WEATHER/9703/05/flood/index.html* (viewed November 11, 1998)

CNN Weather. 1997c (March 7). Ohio River crests in Louisville. *http://cnn.com/WEATHER/9703/07/flood.update/index.html* (viewed November 11, 1998)

CNN Weather. 1997d (March 10). Ohio River flood danger far from over. *http://cnn.com/WEATHER/9703/10/flooding/index.html* (viewed November 11, 1998)

CNN World. 1997a. Egypt launches controversial Peace Canal Project. *http://cnn.com/WORLD/9701/09/egypt.canal/index.html* (viewed November 11, 1998)

CNN World. 1997b (February 5). Namibia's water shortage threatens African oasis. *http://cnn.com/WORLD/9702/05/botswana/index.html* (viewed November 11, 1998)

CNN World. 1997c (June 26). Egypt looks to desert to ease overcrowding. *http://cnn.com/WORLD/9706/26/egypt.population* (viewed November 11, 1998)

CNN World. 1998 (May 25). Uganda's plan to dam Nile sparks controversy. *http://cnn.com/WORLD/africa/9805/25/uganda.nile.controversy* (viewed November 11, 1998)

Dartmouth Flood Observatory. 1998. *http://www.dartmouth.edu/artsci/geog/floods* (viewed November 11, 1998)

Egypt Online State Information Service. 1997a (April 1). Mubarak reviews with experts final preparations for Zayed Canal Project. *http://www.sis.gov.eg/online/html/ol050197.htm* (viewed November 11, 1998)

Egypt Online State Information Service. 1997b (September 1). A new delta west of the Nile Valley: project of the century. *http://www.us.sis.gov.eg/calendar/html/cl090197.htm* (viewed November 11, 1998)

Egypt Online State Information Service. 1998. Al-wadi al Faregh (Empty Valley): a new green spot amidst barren desert. *http://www.sis.gov.eg/online/html* (viewed November 11, 1998)

Fels, E., and Keller, R. 1973. World register on man-made lakes. In Ackermann, W. C., White, G. F., Wothington, E. B., and Ivens, J. L., eds., *Man-Made Lakes: Their Problems and Environmental Effects*. Washington, DC: American Geophysical Union, pp. 43–49.

Frederick, K. D. 1996. Water as a source of international conflict. *Resources,* 123, Spring. Resources for the future. *http://www.rff.org/resources_articles/files/waterwar.htm* (viewed November 11, 1998)

Gardner-Outlaw, T., and Engelman, R. 1997. Sustaining water, easing scarcity: a second update. *Population Action International.* *http://www.populationaction.org/why_pop/water/water-toc.htm* (viewed November 11, 1998)

Glantz, M. 1994. *Drought Follows the Plow.* Cambridge: Cambridge University Press, 197 pp.

Glazovskiy, N. F. 1995. The Aral' Sea basin. In Kasperson, J. X., Kasperson, R. E., and Turner, B. L., II, eds., *Regions at Risk: Comparisons of Threatened Environments.* Tokyo: United Nations University Press, 588 pp.

Gleick, P. H. 1993. *Water in Crisis: A Guide to the World's Fresh Water Resources.* New York: Oxford University Press, 473 pp.

Goolsby, D. A., and Battaglin, W. A. 1995. Effects of episodic events on the transport of nutrients to the Gulf of Mexico. *Proceedings, First Gulf of Mexico Hypoxia Management Conference,* December 5–6, Kenner, LA. Available at: *http://pelican.gmpo.gov/nutrient/front.html* (viewed November 10, 1998)

Gore, A. 1992. *Earth in the Balance: Ecology and the Human Spirit.* New York: Houghton Mifflin, 407 pp.

International Boundary Water Commission. 1998. *http://www.ibwc.state.gov* (viewed November 10, 1998)

Liniger, H. 1995. *Endangered Water: A Global Overview of Degradation, Conflicts and Strategies for Improvement.* Berne, Switzerland: Group for Development and Environment, University of Berne, 117 pp.

L'vovich, M. I. 1979. In Nace, R. L., ed., *World Water Resources and Their Future.* Special Publication, translation. Washington, DC: American Geophysical Union, 416 pp.

MacLeod, N. H., Schuber, J. S., and Anaejionu, P. 1977. Report on the Skylab 4 African Drought and Arid Lands Experiment. In *Skylab Explores the Earth*. NASA SP-380. Washington, DC: NASA, pp. 263–286.

McCaffrey, S. C. 1993. Water, politics and international law. In Gleick, P. H., ed., *Water in Crisis: A Guide to the World's Fresh Water Resources*. New York: Oxford University Press, pp. 92–113.

Micklin, P. 1988. Dessication of the Aral' Sea: a water management disaster in the Soviet Union. *Science,* 241:1170–1176.

Micklin, P. 1991. *The Water Management Crisis in the Soviet Central Asia*. Carl Beck Papers, No. 905. Pittsburgh, PA: Center for Russia and East European Studies, University of Pittsburgh, 120 pp.

Micklin, P. 1993 (April). The shrinking Aral' Sea. *Geotimes,* pp. 14–18.

Middlet, N. 1996. The Aral' Sea tragedy. *Geography Review,* 9(3):7–8.

Myllylä, S. 1995. Cairo: a mega-city and its water resources. *Proceedings, 3rd Nordic Conference on Middle Eastern Studies: Ethnic Encounter and Culture Change,* Joensuu, Finland, June 19–22. Available at: *http://www.hf-fak.uib.no/institutter/smi/paj/Myllyla.html* (viewed November 11, 1998)

National Oceanic and Atmospheric Administration. 1996a. NOAA Special Climate Summary, 93/2. Drought in the southern plains and the southwest. *http://nic.fb4.noaa.gov:80/products/special_summaries/96_2* (viewed November 11, 1998)

National Oceanic and Atmospheric Administration. 1996b. NOAA Special Climate Summary, 93/3. Update on drought in the southern plains and the southwest. *http://nic.fb4.noaa.gov:80/products/special_summaries/96_3* (viewed November 11, 1998)

National Park Service. 1998. Amistad National Recreation Area. *http://www.nps.gov/amis* (viewed November 11, 1998)

NIH Image home page. 1998. *http://rsb.info.nih.gov/nih-image* (viewed February 4, 1999)

Office of Earth Sciences Database. 1999. Office of Earth Sciences database of photographic information and images. *http://eol.jsc.nasa.gov/sseop/index.html* (viewed February 15, 1999)

Postel, S. 1992. *Last Oasis: Facing Water Scarcity*. New York: Worldwatch Institute, 239 pp.

Raheja, P. C. 1973. Lake Nasser. In Ackermann, W. C., White, G. F., Wothington, E. B., and Ivens, J. L., eds., *Man-Made Lakes: Their Problems and Environmental Effects*. Washington, DC: American Geophysical Union, pp. 234–245.

Showstack, R. 1997. Gulf of Mexico hypoxic zone stirs policy discussions. *Eos, Transactions of the American Geophysical Union,* 78:478.

Turkish Ministry of Foreign Affairs Web site. 1998. G.A.P. project. *http://www.mfa.gov.tr/grupc/gap.htm* (viewed November 11, 1998)

Weekly Mail and Guardian. 1996a. Plan could turn Okavango to dust. *http://wn.apc.org/wmail/issues/961129/NEWS66.html* (viewed November 11, 1998)

Weekly Mail and Guardian. 1996b. Namibia almost certain to drain Okavango. *http://www.mg.co.za/mg/news/96dec1/06dec-namibiadrain.html* (viewed November 11, 1998)

Werner, L. 1997. Revising a river. *Earth,* June, pp. 16–17.

Chapter

5

The 1997–1998 El Niño: Images of Floods and Drought

Cynthia A. Evans, Julie A. Robinson, M. Justin Wilkinson, Susan Runco, Patricia W. Dickerson, David L. Amsbury, and Kamlesh P. Lulla

Office of Earth Sciences
NASA Johnson Space Center
Houston, Texas USA

ABSTRACT

Astronauts and cosmonauts on the *Mir* space station had an unprecedented opportunity to photograph and present a global view of the effects of the 1997–1998 El Niño from July 1997 through May 1998. Their photographs document both the local effects and the supraregional extent of weather anomalies in areas where El Niño conditions have traditionally been reported, such as Southeast Asia, Australia, and northern South America. Significantly, effects of weather anomalies were also documented in regions far from the equatorial Pacific, such as southern Canada/northeastern United States, southern Africa, and northeastern Africa. Progressive drought in Southeast Asia, Australia, the Andean Altiplano, northern Brazil, and South Africa is expressed in lower lake, reservoir, and river levels and increased biomass burning. Abnormally high levels of rainfall and snowpack in typically arid regions of the southwestern United States and northeastern Africa are reflected in unusually green vegetation and higher lake and river levels. Minimal sea ice in the Gulf of St. Lawrence attested to the unusually warm winter there.

5.1 INTRODUCTION

The predominant weather event of late 1997 and early 1998 was the El Niño/Southern Oscillation. In the popular news (e.g., CNN On-Line, 1997; National Oceanic and Atmospheric Administration, 1998a) was extensive coverage of the fires and smoke (Southeast Asia, Australia, Brazil, and Mexico) as well as of floods and storms (Ecuador and Peru, Argentina and Uruguay, western United States, and eastern Africa). Unprecedented types and amounts of data were available through the Internet and scientific journals, including animations and daily maps of winds, sea surface temperatures and temperature deviations, sea levels and sea-level variations, and composite maps of precipitation and temperature deviations around the globe. Shuttle–*Mir* astronauts brought back a continuum of data over many of the areas severely affected by the 1997–1998 El Niño. In this chapter we present a synoptic view, through the photographs taken by the astronauts and cosmonauts, of regions of the world that were affected by the 1997–1998 El Niño event. A full tabulation of the photographs related to El Niño is provided in Table 5.1. Here we include views of flooding in Argentina; extensive smoke palls over Brazil, Southeast Asia, and Central America; fluctuating lake levels in the Andean Altiplano of Bolivia and Chile and in central Australia; lush vegetation in the normally arid regions of northeast Africa and California; and minimal winter ice cover in the northeastern United States and southern Canada. To help interpret the photographs, we combine them with other data on regional temperature and precipitation anomalies. The photographs taken from *Mir* late in 1997 and early in 1998 are also contrasted with baseline images from the same regions taken before the 1997–1998 El Niño. The sets of photographs, which can be interpreted without extensive processing, provide a new perspective and a better understanding the impact of short-term climate change on the lives and economies of several regions around the world.

5.2 BACKGROUND

The El Niño–Southern Oscillation (ENSO) is defined on the basis of anomalous atmospheric pressures, elevated sea surface temperatures, and changes in sea surface heights in the equatorial Pacific region (Cane, 1986; Wallace and Vogel, 1994; National Oceanic and Atmospheric Administration, 1998b, c). Atmosphere–ocean coupling causes pulses of tropical Pacific water to travel eastward and pool along the coast of South America in the vicinity of Peru and Ecuador (e.g., Cane, 1986; Wallace and Vogel, 1994; Fleet Numerical Meteorology and Oceanography Center, 1998). The redistribution of heat resulting from the eastward movement of warm Pacific water disrupts global weather patterns for several months, typically from about October through March (Glantz, 1991, 1996; Wallace and Vogel, 1994; Climate Prediction Center, 1998; International Research Institute for Climate Prediction, 1998; National Oceanic and Atmospheric Administration, 1998b, d).

Perhaps the most visible impacts of an ENSO result from the extreme patterns of precipitation that occur worldwide. Figure 5.1 (see color insert) is a map of sea surface temperature anomalies from early December 1997. The figure also summarizes the regions of floods and droughts from November 1997 through January 1998 (from Climate Prediction Center, 1998) as well as the locations of images featured in this chapter.

For the first time, astronauts and cosmonauts had the opportunity to fully observe and document the global impact of an entire ENSO event as it developed and ebbed. This continuum of imagery, which started in March 1996, included baseline pre-El

TABLE 5.1 Photographs Related to El Niño

Topic	Date	Photo Numbers
South America		
Water levels in Lake Poopó, Salar Coipasa	October 1996	NM22-720-009
Water levels in Lake Poopó, Salar Coipasa	October 23, 1996	NM22-725-021
Water levels in Lake Poopó	December 1, 1996	NM22-737-051
Water levels in Lake Poopó, Salar Coipasa	December 1, 1996	NM22-737-053
Water levels in Lake Poopó, Salar Coipasa	March 18, 1997	NM23-714-627
Water levels in Lake Poopó, Salar Coipasa	August 1997	NASA5-704-053
Water levels in Lake Poopó	August 1997	NASA5-705-085
Water levels in Lake Poopó	May 16, 1998	NASA7-726-035 and 036
Water levels in Salar Coipasa	April 11, 1997	NM23-724-630
Water levels in Salars Coipasa and Uyuní	August 1997	NASA5-703-096
Water levels in Lake Poopó, Salar Coipasa	November 16, 1997	NASA6-710-082
Salar Coipasa	November 2, 1996	NM22-729-001
Lake Titicaca	August 1997	NASA5-705-082
Titicaca coast, detail	November 2, 1996	NM22-729-002, 003, and 004
Lake Titicaca	October 1996	NM22-720-008
Lake Titicaca	May 16, 1998	NASA7-726-029 to 033
Lake Titicaca, comparative view	June 1985	STS51G-37-028
Lagunas Colorada and Verde, Altiplano	August 1997	NASA5-704-062
Lagunas Colorada and Verde, Altiplano	October 23, 1996	NM22-725-020
Rio Ucayuli, Peru	August 1997	NASA5-703-002
Rio Uyucali and Rio Pachetea, Peru (11°N, 74°W)	October 12, 1996	NM22-722-038
Pisco, Peru	October 12, 1996	NM22-722-034
Coastal Peru north of Lima	October 12, 1996	NM22-722-035
Coastal Peru at Chile border	December 1, 1996	NM22-737-048
Lakes inland from Lima (Lake Dejunin, Rio Pacheta, Ucayali)	October 12, 1996	NM22-722-033, 036, and 037
Lakes inland from Lima	October 26, 1996	NM22-726-086
Guayaquil in sunglint	November 15, 1996	NM22-734-098
Coastal Peru (Huacho)	April 11, 1998	NASA7-717-050
Pisco, Peru	April 1, 1998	NASA7-715-060
Coastal Colombia (2°N, 75°W)	May 8, 1998	NASA7-722-050
Water levels in Lakes Poopó and Titicaca, smoke pall from Amazon Basin	August 1997	NASA5-705-080
Amazon haze	August 1997	NASA5-706-135
Amazon haze	August 1997	NASA5-704-041
Smoke over Acre state	January 1998	STS089-715-068 to 070
Brazilian Reservoir	October 12, 1996	NM22-726-086
Rio Grande	December 1, 1996	NM22-737-054
Rio Bermejo	December 1, 1996	NM22-737-057
Reservoir, São Francisco (19.3°S, 47°W)	November 25, 1996	NM22-737-009
Manaus and Amazon below Manaus	December 21, 1996	NM22-745-007 to 010
Rio Paraná (dry)	December 1, 1996	NM22-737-064 and 065
Rio Paraná in flood	November 13, 1997	NASA6-710-077 and 078
Rio Paraná reservoirs in flood	April 12, 1998	NASA7-718-027 to 029
Northern Venezuela, dry	October 28, 1997	NASA6-709-086
Gulf of Venezuela, Maracaibo	January 1998	NASA6-712-029 to 035
Rio Magdalena Mouth	January 10, 1998	NASA6-712-047

(Continues)

TABLE 5.1 *(Continued)*

Topic	Date	Photo Numbers
North Africa		
Somalia coast (1°N, 44°E)	March 1996	NM21-727-006
Somalia coast	November 16, 1996	NM22-736-023 to 027
Somalia coast (1°N, 44°E)	January 1, 1998	NASA6-708-056
Somalia coast	February 27, 1998	NASA7-707-084
Somalia coast	April 11, 1998	NASA7-716-083
Somalia coast, 1982	November 1982	STS005-40-1196
Somalia coast, 1989	March 1989	STS029-76-081
Erta Ale and lakes around Erta Ale	December 1997	NASA6-711-011
Lake Edward	April 17, 1998	NASA7-721-007
Lake Albert	April 17, 1998	NASA7-721-010
Lake Natron	October 17, 1996	NM22-724-015
Lake Natron	July 1997	NASA5-701-029
Lake Chad	December 23, 1996	NM22-741-041 and 042
South Africa		
Kalahari Desert smoke palls	August 1997	NASA5-706-195
Green vegetation, southwest South Africa	July 28, 1997	NASA5-705-055
Asia		
Sumatra fires	September 27, 1997	STS0865087 to 5098
Clearings, Kalimantan	October 16, 1996	NM22-723-006 to 011
Clearings in Kalimantan (1.2°S, 102.5°E)	December 21, 1996	NM22-740-041
Coastal Kalimantan fires and clearings (0.5°N, 118°E)	November 1997	STS087-724-056 and 058
Coastal Kalimantan clearings (3.8°N, 112.5°E)	November 1997	STS087-724-040
Central Luzon in sunglint	November 13, 1996	NM22-734-018
Northwestern India, Rann of Kutch (dry)	November 16, 1996	NM22-735-076 and 077
Northwestern India, Rann of Kutch, flooded		
Mekong River, Cambodia, dry	November 1997	STS087-768-084
Mekong River, Cambodia, dry	January 2, 1998	NASA6-708-086
Papua New Guinea, smoke over Owen Stanley Range	October 27, 1997	NASA6-709-064 to 067
North Vietnam, burning	March 17, 1998	NASA7-712-048 to 052
Australia		
Lake Eyre	November 1997	STS087-778-022
Lake Eyre	January 1998	STS089-717-055
Lake Eyre	July 1997	STS094-748-083
Lake Eyre	March 8, 1998	NASA7-710-075 to 077
Lake Eyre	March 17, 1998	NASA7-712-070
Lake Eyre	March 25, 1998	NASA7-714-012 to 014
Lake Eyre	May 6, 1998	NASA7-722-043
Lake Neal and Lake Amadeus	January 1998	STS089-717-042
Lake Neal and Lake Amadeus	September 1997	STS086-715-079
Lake Amadeus	January 1998	STS089-717-045
Lake Amadeus	September 1997	STS086-715-077
Australian lakes	March 8, 1998	NASA7-710-047, 048, 055, and 075 to 085
Australian lakes	March 9, 1998	NASA7-709-065
Australian lakes	April 11, 1998	NASA7-717-047

TABLE 5.1 *(Continued)*

Topic	Date	Photo Numbers
Australian fires	November 1997	STS087-708-031
Australian fires, south tip	March 9, 1998	NASA7-709-086 to 088
Channel country	March 17, 1998	NASA7-712-083
Channel country	March 25, 1998	NASA7-714-040 to 045
Southern Australian lakes	March 25, 1998	NASA7-714-006 to 008 and 025-034
North America		
California greening	March 8, 1998	NASA7-709-029 to 034
Great Salt Lake	October 23, 1996	NM22-725-037
Bering Sea plankton	August 8, 1997	STS085-706-063
Ice pack in Great Lakes, pre-El Niño, Lake Erie	February 20, 1997	NM22-773-083
Ice pack in Great Lakes, pre-El Niño, Lake Superior	February 23, 1997	NM22-777-041
Ice pack in Lakes Erie and Ontario	February 25, 1998	NASA7-707-049
Ice pack in St. Lawrence Seaway, pre-El Niño	February 1997	NM22-778-070 to 100; NM22-777-092
1997–1998 ice pack, St. Lawrence Seaway	January 4, 1998	NASA6-712-009 and 010
Gulf of St. Lawrence	February 28, 1998	NASA7-706-078
Bay of Fundy	February 28, 1998	NASA7-706-079
Nova Scotia coast, no ice	April 15, 1998	NASA7-718-045
Isles de la Madeleine, St. Lawrence mouth	December 29, 1997	NASA6-708-003
Mexican smoke	April 9, 1998	NASA7-713-061 to 065
Smoke over Yucatan	April 15, 1998	NASA7-719-039
Mexican smoke	April 17, 1998	NASA7-721-047 to 052
Mexican smoke	May 10, 1998	NASA7-723-006 and 007
Mexican smoke	May 19, 1998	NASA7-725-022 to 026
Mexican smoke	May 19, 1998	NASA7-725-038 to 057
Smoke over Costa Rica	May 22, 1998	NASA7-726-004 to 009
Smoke over Central America	May 16, 1998	NASA7-726-09 to 020

Niño conditions (water levels and vegetation distribution) leading up to the record-breaking 1997–1998 El Niño. The data/photographs are rich visual portrayals of places that are classically effected by El Niño events. The 1996–1998 images have been correlated with imagery from the same regions taken during earlier missions to help define the spatial and temporal scale of well-known events such as droughts and fires in Australia, Indonesia, and New Guinea; regional droughts and floods in various parts of South America; and the "greening" of the U.S. southwest, coastal Peru and Ecuador, and northeastern Africa. Some of the best illustrations of the impact of the severe weather include water-level fluctuations (dropping water levels in lakes in Australia and the Andean Altiplano and drier than usual rivers in the Amazon, and rising water levels in the southwestern United States and northeastern African lakes) and large smoke palls from biomass burning in Southeast Asia, Australia, South America, and South Africa. The photographs also help to explain the connections between the El Niño phenomenon and its major economic impacts, such as devastating droughts that result in large fires and crop failures, and equally devastating floods, which resulted in crop failures, landslides, and the spread of waterborne disease.

5.3 SOUTH AMERICA

The weather term *El Niño* is related to the occurrence of warm waters off coastal Peru around the Christmas season (e.g., Cane, 1986) (Figure 5.1). This pattern of unusual warm water off the Peruvian coast was prevalent from July 1997 through March 1998 (Fleet Numerical Meteorology and Oceanography Center, 1998). The best-known economic impact of these warm, nutrient-poor waters is the decline of the Peruvian anchovy industry. Both Chile and Peru levied a ban on fishing in September 1997, and by November, the Colombian fish catch had already dropped by more than 20% (Duffy and Bryant, 1997). But the precipitation pattern over the entire South American continent was also disrupted. During El Niño events, much of Peru and Ecuador typically receive heavy rainfall and experience flooding, and the Paraná River drainage in Argentina and Uruguay also experiences heavier than normal rainfall. But at the same time, much of northeastern South America and the high central Andes of Bolivia and northern Chile experience drought (Glantz, 1991, 1996; Pearce, 1994; Wallace and Vogel 1994; Climate Prediction Center, 1998; National Drought Mitigation Center, 1998).

In January and February 1998, heavy floods in Peru and Ecuador resulted in catastrophic floods and landslides—national emergencies were declared in Peru, Chile, Ecuador, Bolivia, and Colombia (CNN Specials, 1998; CNN World News, 1998; Earthweek, 1998c; Pan-American Health Organization, 1998). Floods also ravaged Uruguay and northern Argentina, including the Paraná River drainage (Figure 5.2). At the same time, more than six months of drought in northeastern Brazil fueled brush fires that charred vast acreage (over 56,000 km^2) at unprecedented rates, threatening crops, livestock and resulting in starvation for roughly 40,000 Indians (Earth Alert, 1998a, b; Food and Agriculture Organization, 1998b, c; Pan-American Health Organization, 1998). These fires came several months after a severe dry season in August–September in the Amazon basin, where burning was also extensive (World Wildlife Fund, 1997; Earthweek, 1998a; National Oceanic and Atmospheric Administration, 1998c).

As the impacts of the 1997–1998 El Niño intensified during the NASA 5 mission (May through September 1997), astronaut Mike Foale was able to observe and record the extensive smoke palls over South America (Figure 5.3). The dry weather over the high Andes resulted in the predictable lowering of water levels in some of the high lakes, which are excellent markers for short-term climate fluctuations. Over the years, the Earth Sciences Office has used water levels in Lago Poopó in Bolivia and water in Salars Coipasa and Uyuní as benchmarks for relative rainfall in the Andean Altiplano. In the previous extended El Niño period from 1990 through 1995 (Pearce, 1994; Trenberth and Hoar, 1996), Lago Poopó evaporated completely. From the mid-1980s through 1990, the water levels were high, declining only after 1991. We have also monitored water levels and swamp vegetation around Lake Titicaca. However, because Lake Titicaca is substantially deeper, the relative amounts of evaporation are more difficult to detect and measure along Titicaca's shoreline.

Figure 5.4 (see the color insert) shows a short time series of photographs that document changes in the water levels in these Andean basins during the most recent El Niño event. The first photo in this figure, NM23-714-627, features Lago Poopó and

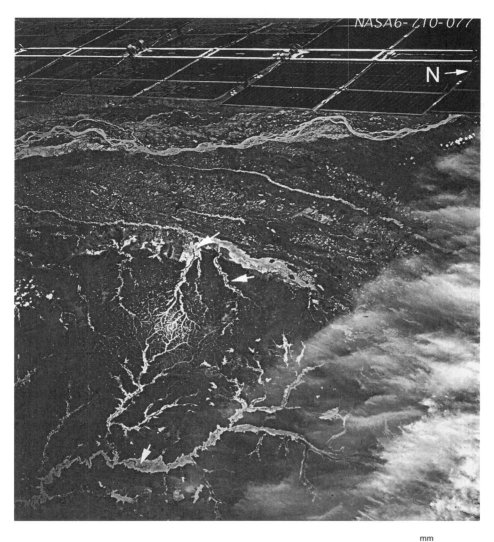

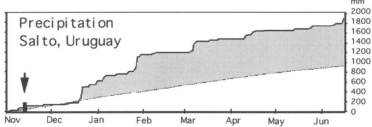

Figure 5.2 Severe flooding in the Paraná drainage of northern Argentina and Uruguay can be seen in this early January 1998 view (NASA photograph NASA6-710-077). The photo looks west over the Rio Meriñay (foreground) and the Paraná (under the solar panel). Sunglint highlights the immense area covered in water—the edges of the normal river drainages can barely be discerned in the larger flooded river valleys (arrows). Salto, Uruguay is just off the edge of the image to the lower left. The graph shows the regional precipitation deviations due to El Niño. The cumulative observed precipitation is given by the heavy solid line, and normal precipitation (cumulative) is depicted by the thin dashed line. The bar indicates when the photo was taken. (Modified from the Climate Prediction Center, 1998.)

August, 1997

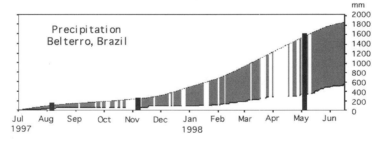

Figure 5.3 Dry season in Brazil. This photograph (NASA photograph NASA5-705-080) looks southeastward over Lake Titicaca in the Peruvian–Bolivian Andes (foreground) and Lago Poopó (Bolivia) toward the smoke-induced haze of Brazilian and Bolivian lowlands. The late July 1997 Amazon smoke pall, the result of biomass burning in the Amazon Basin during the dry season, is the light haze filling the background. The smoke finds its way up valleys from the basin even into the upper reaches of the Andes (center left of the image). The dry season in parts of the Amazon basin was unusually severe in 1997, which has been attributed to El Niño. The graph shows the regional precipitation deviations due to El Niño. The cumulative observed precipitation is given by the heavy solid line, and normal precipitation (cumulative) is depicted by the thin dashed line. The bars indicate when this photo and the photos in Figure 5.4 were taken. (Modified from the Climate Prediction Center, 1998.)

Salar de Coipasa during the flooding event in March 1997, which filled the basins for the first time since 1990. The heavy precipitation provided a benchmark for the subsequent drought and water drawdown experienced by the region. Shortly after April 1997, the water in these basins began to evaporate. Figure 5.5 is a compilation of water levels in Lago Poopó and Salar de Coipasa created from a time series of photographs taken by astronauts since 1988. The water area in the two lake beds was mapped from each image onto the aircraft Operational Navigation Chart (ONC)-series base map, from which the area was calculated using *NIH Image* software. The water areas are graphed versus time to show the coincidence of water-level drops in these lakes, specifically with El Niño events.

5.4 SOUTHEAST ASIA

One of the earliest and most newsworthy events of the 1997–1998 El Niño was the drought in Southeast Asia, and the smoke pall that was produced by hundreds of fires raging out of control on Sumatra, Kalimantan, and other parts of the region. The resultant smoke pall covered all of Southeast Asia for more than a month, and resulted in widespread health problems and local catastrophes such as airport closures, airline crashes, and boat collisions (Center for Remote Imaging, Sensing and Processing, 1997, 1998; Showstack, 1997; Swinbanks, 1997; World Wildlife Fund,

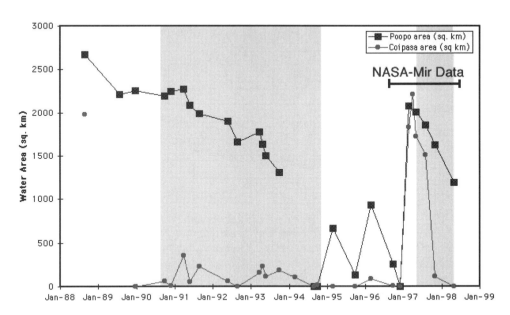

Figure 5.5 Graph of water areas in Lago Poopó and Salar de Coipasa, Bolivia. The data were mapped from photographs taken by astronauts and cosmonauts. The gaps represent periods for which no data are available. The El Niño events are indicated by shading in the background.

1997; BBC News, 1998; Food and Agriculture Organization, 1998b, c; Haze Online Report, 1998; Meteorological Service, Singapore, 1998). Both the areal extent of the smoke and the opacity of the smoke pall as viewed from space were impressive. Figure 5.6 shows an edge of the large smoke pall covering Sumatra in September 1997. Other imagery taken in October 1997 during NASA 6 shows equally thick smoke palls over Papua New Guinea (see also Fig 7.6). Drought and fires in the western Pacific region are well-known El Niño occurrences (Glantz, 1996), and El Niño–related smoke has been observed by astronauts on earlier Shuttle missions. For example, fires in Indonesia, New Guinea, and Australia were photographed extensively by Space Shuttle crews in September and October 1994 (e.g., STS068-269-045 and STS068-261-062).

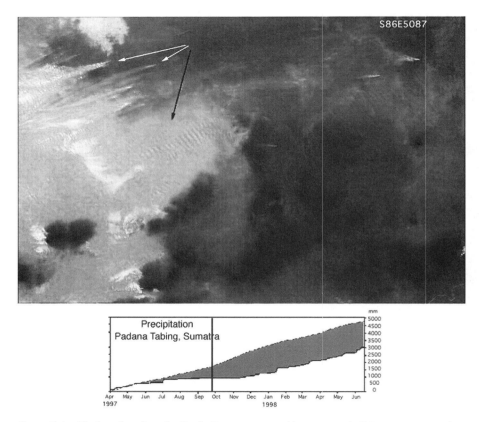

Figure 5.6 The first photo from the Shuttle-*Mir* program to visibly portray early El Niño impacts was taken during the STS-86 mission in September 1997, as Dave Wolf was exchanging places with Mike Foale (NASA ESC image S86E5087). The Electronic Still Camera on the Shuttle, controlled by middle school students involved in the KidSat project (KidSat, 1997), was used to create a photomosaic of the smoke pall covering southern Sumatra. This is one of these images showing a corner of the smoke pall covering the southern tip of Sumatra. The white arrows mark some of the individual smoke plumes, the dark arrow points to dense smoke with gravity waves, which could be mistaken for cloud. The graph shows the regional precipitation deviations due to El Niño. The cumulative observed precipitation is given by the heavy solid line, and normal precipitation (cumulative) is depicted by the thin dashed line. The bar indicates when the photo was taken. (Modified from the Climate Prediction Center, 1998.) See also color insert for Figure 7.6.

5.5 AUSTRALIA

From September through mid-December, much of Australia was exceptionally hot and dry, and out-of-control fires produced significant smoke palls. NASA 6 crew member David Wolf was able to take some oblique photographs of the smoke. In late December and January 1998, a persistent tropical storm dumped rains on the northeastern part of the continent, but much of the central and southeast regions remained hot and dry (Climate Prediction Center, 1998; Earthweek, 1998b). As for the high plains of the central Andes, we have used lake levels in south-central Australia as visual markers of the relative amounts of precipitation. Early in 1997, Lake Eyre, the lowest part of the Australian continent, accumulated water for the first time in several years. Our time series starts in July 1997. The water levels dropped rapidly from that time, were much lower by January 1998, and were gone completely by April 1998 (Figure 5.7; see the color insert). Drying of Lake Eyre has been associated with previous El Niño events, and flooding of Eyre is thought to follow after the end of an El Niño event (Kotwicki, 1986). During the last major El Niño in 1982–1983, Lake Eyre dried up, but then flooded in 1984 (Allan, 1985; Kotwicki, 1986). We anticipate that future astronauts will document post–El Niño flooding in Lake Eyre within a year.

5.6 AFRICA

Another catastrophic event that has been related to the 1997–1998 El Niño event was the widespread flooding in northeastern Africa during December 1997 and January 1998. Prior to this El Niño event, forecasts indicated that wetter than normal conditions would be concentrated over the Kenya–Tanzania region (Climate Prediction Center, 1998). The countries of Somalia, Ethiopia, Kenya, and Tanzania received 350 to 1300 mm of rain within a few weeks, between 5 and 60 times the normal amount of rainfall for this typically arid region (Climate Prediction Center, 1998; ReliefWeb, 1998). The floods washed out villages and roads, making relief work nearly impossible. In addition to shortages of food, water, and medical supplies, the destruction of crops and livestock, and new epidemics of waterborne diseases, affected the region (Earthweek, 1998a; Food and Agriculture Organization, 1998a–c; ReliefWeb, 1998). Very few climatology data are collected in Somalia (Climate Prediction Center, 1998), but Figure 5.8 (see the color insert) depicts the effects of the unusually heavy rains on Somalian coastal vegetation. The two views pictured in Figure 5.8 were taken roughly two years apart along the Somalia coast, near the town of Barraawe south of Mogadishu. Although the flooded rivers and roads are not visible in the figure, the greening and apparent lushness of this arid region in January 1998 stands in contrast to the earlier, more typical view, and serves to illustrate the large changes in precipitation. Increased rainfall over Somalia also occurred during the 1982–1983 El Niño event. Images from an earlier Space Shuttle flight (e.g., STS005-40-1196, November 1982) show similar greening along the Somalia coast.

The antithesis of an El Niño, the Pacific cold episode (La Niña), began in May 1998 and is forecast to continue through the winter of 1998–1999 (Climate Prediction Center, April 1998). A March 1989 photograph of the same region documents very dry conditions on the Somalia coast during the last cold episode in 1989 (STS029-76-081). This region might expect similarly dry conditions during the

1998–1999 winter. Future imagery from space will verify whether the extremely dry La Niña conditions occur in Somalia during the 1998–1999 winter.

In southern Africa, a different picture emerged during the 1997–1998 El Niño. Instead of persistent rains, much of southern Africa experienced dry weather (Climate Prediction Center, 1998; National Drought Mitigation Center, 1998). In late July–early August 1997, NASA 5 crew member Mike Foale documented impressive savanna fires across the whole of the Kalahari Desert. Although it is difficult to assess unequivocally whether the smoke was more extensive than usual, our sense was that there was a significant increase in the resultant smoke palls.

5.7 NORTH AMERICA

North America was also affected by anomalous weather attributed to the 1997–1998 El Niño. Notably, heavy rains pummeled California and the southeastern United States, while the central Rocky Mountains received less snow than usual early in the winter. The heavy precipitation to the California coastal region, the Central Valley, and the Sierra Nevada created remarkably green vegetation along the coast and along the foothills of the mountains. Figure 5.9 (see the color insert) is a March 1998 view of central California showing the unusual green color of the landscape.

The northeastern United States enjoyed a remarkably mild winter. Temperatures from January to March were 3 to 9°F warmer than usual (Climate Prediction Center, 1998; Northeast Regional Climate Center, 1998). The ice pack in the Great Lakes never really developed (Figure 5.10), and ice booms were removed by early March

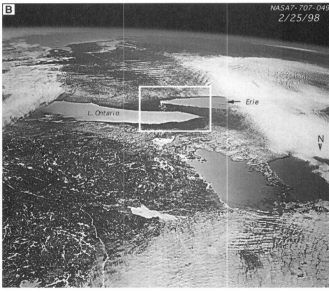

Figure 5.10 These two views compare the ice on Lake Erie on (*A*) February 20, 1997 and (*B*) February 25, 1998 (NASA photographs NM22-773-083 and NASA7-707-049). View (*B*) looks south over the lakes toward the Chesapeake on the horizon. Sunglint highlights the water. The box shows the approximate area covered in view (*A*). Despite regional snow cover, no ice covered the eastern Great Lakes in February 1998 when view (*B*) was taken.

(New York Power Authority, 1998). This stands in contrast to the 1996–1997 winter across the central and eastern United States and Canada, when NASA 4 crew member Jerry Linenger was able to photograph a substantial ice pack in the Great Lakes, Hudson Bay, and the St. Lawrence seaway into late March 1997. In fact, both the NASA 2 and NASA 4 crews saw ice in Hudson Bay through July of 1996 and 1997, respectively. To illustrate the impact of the average temperature difference between 1997 and 1998, we compare the late February ice cover in Lake Erie in 1997 and 1998 (Figure 5.10) and the 1997 and 1998 winter sea ice cover in the St. Lawrence Seaway in Figure 5.11.

In Mexico and much of Central America, extreme drought supported extensive wild fires that burned out of control for more than two months. In April and May 1998, the NASA 7 crew photographed the smoke palls repeatedly (Figure 5.12). At the height of the burning, the smoke covered almost all of Central America, Florida, Texas, and extended north to Canada. Table 5.1 includes the most significant views, and the smoke is discussed in more detail in Chapter 7.

5.8 CONCLUSION

The photographs presented here are a global survey of the 1997–1998 El Niño impacts and allow a visual assessment of the effects of this recurring, extreme weather

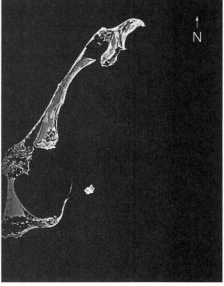

Figure 5.11 A visual marker for the unusually mild winter over the northern United States and southern Canada was the lack of an extensive ice cover on the St. Lawrence Seaway. An early January 1998 view of the Isles de la Madeleine (NASA photograph NASA6-708-003) usually icebound by midwinter, contrasts with a February 1997 regional view (NASA photograph NM22-778-081), which shows extensive ice cover in the southeastern Gulf of St. Lawrence between Newfoundland (foreground) and Nova Scotia (left of center). The arrow points to the Isles de la Madeleine.

Figure 5.12 Fires burning out of control in central and southern Mexico and Central America created a regional smoke pall that covered much of Central America and extended into the Pacific and Caribbean region and as far north as Canada. This view (NASA photograph NASA7-726-009) was taken near the height of the fires in mid-May 1998. It looks north from Nicaragua in the foreground toward Mexico at the top; smoke extends to the horizon.

pattern. The images documenting the effects of severe El Niño–induced weather are quite easily interpreted by the general public and may be a valuable tool for discussing predicted effects of global climate changes. In particular, these images can serve as illustrations for the magnitude of possible climate changes under global warming scenarios. But more important, these data, which cover a complete El Niño event, become part of the larger image database. The photo time series over regions usually affected by El Niño, like the high Andean plain and central Australia, is extensive and allows for comparisons to be made and quantifiable trends to be calculated.

ACKNOWLEDGMENTS

We gratefully acknowledge the crew members who obtained the images—despite difficulties while on orbit, they provided us with this unprecedented data set. The authors also thanks Leslie Upchurch, Marco Lozano, and Joe Caruana for input and comments. Finally, Mike Helfert's continual stream of information and of comments and encouragement in the review process greatly improved the manuscript.

REFERENCES

Allen, R. 1985. *The Australian Summer Monsoon: Teleconnections and Flooding in the Lake Eyre Basin.* Monograph 2. Royal Geographical Society of Australia, South Australia Branch.

BBC News. 1998 (February 25). Haze: who starts the fires. *http://news.bbc.co.uk/hi/english/world/analysis/newsid_59000/59880.stm* (viewed September 21, 1998)

Cane, M. 1986. El Niño. *Annual Review of Earth Planetary Sciences*, 14:43–70.

Center for Remote Imaging, Sensing, and Processing, National University of Singapore. 1997. *http://www.crisp.nus.edu.sg* (viewed September 21, 1998)

Center for Remote Imaging, Sensing, and Processing, National University of Singapore. 1998. SPOT satellite images of forest/plantation fires and smoke haze over Southeast Asia. *http://137.132.32.52/forest_fire/forest_fire.html* (viewed September 21, 1998)

Climate Prediction Center. 1998. CPC: analysis and monitoring products. *http://nic.fb4.noaa.gov:80/products/analysis_monitoring* (viewed September 21, 1998)

CNN On-Line. 1997. El Niño returns. *http://www.cnn.com/SPECIALS/el.nino* (viewed September 21, 1998)

CNN Specials. 1998. El Niño in Peru: a new lake and a wake of destruction. *http://www.cnn.com/SPECIALS/el.nino/peru* (viewed September 21, 1998)

CNN World News. 1998 (January 29). *http://www.cnn.com/WORLD/9801/29/peru.mudslide* (viewed September 21, 1998)

Duffy, D., and Bryant, P. J. 1997. The 1997 El Niño/Southern Oscillation. ENSO 97–98. *http://darwin.bio.uci.edu/~sustain/ENSO.html* (viewed September 21, 1998)

Earth Alert. 1998a (March 9). *http://www.discovery.com/news/earthalert/980309/index/drought_index.html* (viewed September 21, 1998)

Earth Alert. 1998b (March 16). *http://www.discovery.com/news/earthalert/980316/index/fires_index.htm* (viewed September 21, 1998)

Earthweek. 1998a (January 23). *http://www.earthweek.com/history/012398* (viewed September 21, 1998)

Earthweek. 1998b (February 6). *http://www.earthweek.com/history/020698* (viewed September 21, 1998)

Earthweek. 1998c (February 13). *http://www.earthweek.com/history/021398* (viewed September 21, 1998)

Fleet Numerical Meteorology and Oceanography Center. 1998. *http://metoc-u1.fnmoc.navy.mil* (viewed September 21, 1998)

Food and Agricultural Organization. 1998a (February). *Food Outlook*. No. 1. Rome: FAO, United Nations.

Food and Agriculture Organization. 1998b (May). *Food Crops and Shortages*. No. 2. Rome: FAO, United Nations.

Food and Agriculture Organization. 1998c (June). *Food Crops and Shortages*. No. 3. Rome: FAO, United Nations.

Glantz, M. 1991. *ENSO and Climate Change*. Report of workshop held November 4–7, 1991, Bangkok, Thailand. Boulder, CO: National Center for Atmospheric Research.

Glantz, M. 1996. *Currents of Change: El Niño's Impact on Climate and Society*. Cambridge: Cambridge University Press, p. 194.

Haze Online Report. 1998. *http://www.vensara.com/haze* (viewed September 21, 1998)

International Research Institute for Climate Prediction. 1998. ENSO impacts from around the world. *http://iri.ucsd.edu/hot_nino/impacts* (viewed September 21, 1998)

KidSat. 1997. *http://kidsat.jpl.nasa.gov* (viewed September 21, 1998)

Kotwicki, V. 1986. *Floods of Lake Eyre*. Adelaide, South Australia: Engineering and Water Supply Department.

Meteorological Service, Singapore. 1998. Monitoring of smoke and haze and forest fires in Southeast Asia, daily maps. *http://www.gov.sg/metsin/hazed.html* (viewed September 21, 1998)

National Drought Mitigation Center. 1998. *http://enso.unl.edu/ndmc* (viewed September 21, 1998)

National Oceanic and Atmospheric Administration. 1998a. El Niño in the news. *http://www.ogp.noaa.gov/enso/news.html* (viewed September 21, 1998)

National Oceanic and Atmospheric Administration. 1998b. El Niño theme page. *http://www.pmel.noaa.gov/toga-tao/el-nino* (viewed September 21, 1998)

National Oceanic and Atmospheric Administration. 1998c. El Niño–Southern Oscillation home page. *http://www.ogp.noaa.gov/enso* (viewed September 21, 1998)

National Oceanic and Atmospheric Administration. 1998d. El Niño/Southern Oscillation. *http://nic.fb4.noaa.gov/products/analysis_monitoring/ensostuff/index.html* (viewed September 21, 1998)

New York Power Authority. 1998. The Lake Erie–Niagara River ice boom. *http://iceboom.nypa.gov* (viewed September 21, 1998)

Northeast Regional Climate Center, Cornell University. 1998. *http://met-www.cit.cornell.edu/nrcc_home.html* (viewed September 21, 1998)

Pan-American Health Organization. 1998 (April). *Disasters: Preparedness and Mitigation in the Americas*. No. 72. Washington, DC: Pan American Health Organization.

Pearce, F. 1994 (January). Fire and flood greet El Niño's third year. *New Scientist*, 15.

ReliefWeb. 1998. *http://wwwnotes.reliefweb.int* (viewed September 21, 1998)

Showstack, R. 1997. Scientists assess impact of Indonesian fires, *Eos, Transactions of the American Geophysical Union*, 78(44):493.

Swinbanks, D. 1997. Forest fires cause pollution crisis in Asia. *Nature*, 389:321.

Trenberth, K. E., and Hoar, T. J. 1996. The 1990–1995 El Niño–Southern Oscillation event: longest on record. *Geophysical Research Letters*, 23(1):57–60.

Wallace, J., and Vogel, S. 1994 (Spring). El Niño and climate prediction. *Reports to the Nation on Our Changing Planet, UCAR/NOAA*, Vol. 3, p. 25.

World Wildlife Fund, 1997. Rain forest fire crisis: WWF experts report forest fires out of control around the world. *http://www.wwf.org/new/fires/home.htm* (viewed September 21, 1998)

Chapter

6

Imaging Aerosols from Low Earth Orbit: Photographic Results from the Shuttle–*Mir* and Shuttle Programs

M. Justin Wilkinson
Office of Earth Sciences
NASA Johnson Space Center
Houston, Texas USA

J. D. Wheeler and Robert J. Charlson
Department of Atmospheric Sciences
University of Washington
Seattle, Washington USA

Kamlesh P. Lulla
Office of Earth Sciences
NASA Johnson Space Center
Houston, Texas USA

ABSTRACT

Studies of atmospheric aerosols, especially of anthropogenic pollution and dust, are critical for defining parameters in global change models. Photographs taken from low Earth orbit provide additional information to supplement data from ground stations and high-altitude satellite sensors. We use a haze event in the eastern United States as a case study to show the value of the photographs from low Earth orbit when integrated with point data acquired from ground stations. Five examples of probable anthropogenic hazes, some with associated ground data, from northern Italy, the Ukraine, the Red Basin in China, the Nile River delta, and southern Africa show the value of oblique and wide-angle views for capturing aerosol events on film. Through these examples we illustrate the large areas blanketed by haze masses, the high albedo of thick anthropogenic haze, and the well-known topographic control of haze distribution. Four examples from three continents illustrate regional (Tibetan Plateau, Andes Mountains) and intercontinental (Sahara–Europe, China–North America) movement of dust particles in major aerosol events.

6.1 INTRODUCTION

Aerosols, "a suspension of solid or liquid particles in a gas" (National Research Council, 1996:8), can be generated in various ways. "Atmospheric aerosols are ubiquitous and often observable by eye as dust, smoke and haze. Particles . . . range in size from nanometers (nm) to tens of micrometers (μm), that is, from large clusters of molecules to visible flecks of dust. Most of the smallest particles (less than about 0.1 μm) are produced by condensation, either from reactive gases in the atmosphere (e.g., sulfur dioxide) or in high-temperature processes (e.g., fire). Particles larger than about 1 μm are usually produced mechanically (windblown soil, sea spray, etc.)" (National Research Council, 1996:8–9).

The major tropospheric aerosol types are clearly related to the surface over which the host air mass has recently moved—marine aerosol, smoke particulates from regions with biomass burning, industrial haze and desert dust. In this chapter we focus on the latter two types.

The largest natural sources of atmospheric aerosols are major dust-emitting deserts (Sahara, Gobi, etc.) and natural wildfires (grasslands or forest). Anthropogenic aerosols amount to only about 10% of the total atmospheric aerosol load, but particles released by industry and human-induced fires have longer lifetimes in the atmosphere and higher levels of reflectivity, so that their effect on climate forcing is estimated to be nearer to 50% of the total climate forcing by aerosols (National Research Council, 1996). Concerning the anthropogenic group of aerosols, industrial sulfate-producing activities appear to be somewhat more important in generating aerosols than biomass combustion related to land-clearing activities or burning of agricultural wastes (National Research Council, 1996).

6.1.1 Advantages of Astronaut Photographs for Aerosol Documentation

Aerosols can be documented conveniently by astronauts using hand-held cameras despite the fact that hazes are often subtle visual features. Astronaut photographs have special advantages over most other types of satellite images; for example, oblique views can be used to "thicken" the atmosphere by lengthening the line of sight through the aerosol mass. Delicate, otherwise hard-to-view, density boundaries are frequently revealed in this way.

Astronauts in low Earth orbit have the ideal vantage point for synoptic views. Photographic documentation of processes at different scales is achievable through the use of different focal-length lenses and as a result of different orbit altitudes. Low Earth orbit provides a platform from which larger, regional-scale aerosol events can be documented that would otherwise have no visual counterpart. With experience, space-based photographers can begin to recognize anomalous features.

Spacecraft like the Space Shuttle and *Mir* also overfly some of the remotest parts of the planet, where weather stations are nonexistent or are widely dispersed. The typical scale provided by the photographs allows the definition of areas of aerosol events far more precisely than can datasets collected at discrete points on the ground, and act as obvious adjuncts to satellite-based sensors. Data in hand-held photographs can integrate common weather datasets (such as humidity, pressure, and visibility) at higher levels of resolution than are available from weather satellites. For regional studies of aerosol events, photographs taken from low Earth orbit have been referred to as the "cement" which can bind together such data as surface point-source data, aircraft-based data, and satellite imagery (R. Swap, University of Virginia, Charlottesville, 1998, personal communication).

We believe that astronaut photographs are specifically useful to the scientific community and the general public as a tool for visualization. The photographs allow visualization of dynamic smog and dust events in three dimensions, and through time. Such visualization can heighten scientific insight (Hall, 1998; Kaspar, 1998). Furthermore, occasional discoveries are made concerning sources and trajectories of airborne material and the sizes of the masses involved.

Our objectives are (1) to illustrate integration of data from a variety of sources with a photograph taken from low Earth orbit; (2) to provide examples of how anthropogenic pollution events can be observed using the photographs; and (3) to illustrate global dust transportation using the photographs.

6.1.2 Significance of Industrial Aerosols

Interest has recently increased in modeling aerosol masses with high albedos because smog can act as a major reflector of insolation, an effect that results in regional cooling. The scale of this cooling effect is globally averaged at 25% of the total greenhouse-gas warming with a large uncertainty. In fact, absolute cooling of the major industrial regions of the northeastern United States, Western Europe, and East Asia has been recently ascribed to this effect (Charlson et al., 1991; Charlson and Wigley, 1994). Such regionalization of aerosol-induced cooling alters temperature gradients and will have unknown effects on the dynamics of the atmosphere (Charlson and Wigley, 1994).

Charlson and Wigley (1994) used computer-modeled maps to show direct climate forcing by both sulfate aerosols and greenhouse gases. Figure 6.1 was generated from calculations of (1) heat gain in northern summer due to greenhouse gases, and (2) the cooling effect by sulfate aerosol. The global average of heating due to greenhouse gases in this model is calculated to be about 2.2 W m^{-2}, but this is reduced to about 1.7 W m^{-2} when the cooling effect is included (Charlson and Wigley, 1994).

It is particularly significant that this combined forcing is not spatially uniform. A critical result of the model is that the temperature gradient between subtropical and temperate latitudes is increased (Fig. 6.1). These latitudes are the mixing zone where the atmospheric heat-engine exchanges cold polar air masses with hot equatorial air masses.

The meteorological consequences of changes in temperature gradient are as yet unknown, but it is clearly important to understand the role of a gradient change on regional-scale meteorological processes. It seems quite probable that climate change

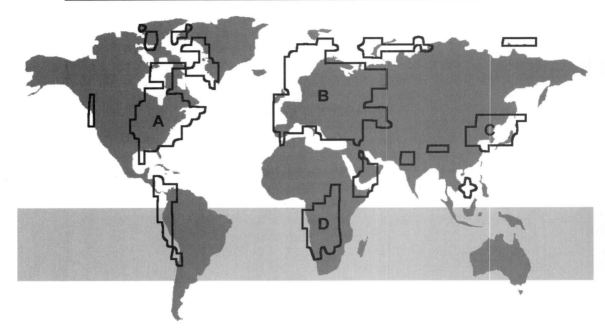

Figure 6.1 Climatic forcing by human activity—July 1993. Models suggest that net cooling may result over the major industrial regions of the world—eastern North America, western Europe, and east Asia—as a result of the reflective effect of anthropogenic haze in the atmosphere (−1 W per square meter and more in the centers of areas A, B, and C). Smoke from biomass burning explains the major southern hemisphere cooling center in Angola (D). Models further suggest an extensive zone a warming south of the equator (shaded box, +1 W per square meter over the oceanic sectors). Cooling in the centers of regions A, B, and C, with warming in much of the shaded box increases temperature gradients (adapted from Kiehl and Briegleb, 1993, and Charlson and Wigley, 1994.)

will not simply involve "global warming" but that regional-scale changes in important factors such as rainfall and windiness will begin to be felt.

The change in temperature gradients occurs in those latitudes where most people live and grow their food: This means that the potential human impacts are great. For example, it has already been suggested that the aerosol blanket over China might have the "teleconnection" effect of decreasing monsoon rainfall in India (references in Mudur, 1995).

Figure 6.1 also shows three regions in which aerosol loadings are calculated to be so heavy that net cooling is probably occurring—the eastern third of North America, Europe, particularly the southern and eastern portions, and East Asia. China is of special interest because it has the most rapidly developing industry and hence increasing aerosol production.

With this level of atmospheric forcing, models now demand among other things, refinement of the spatial and temporal scales of aerosol events. Photographs from low Earth orbit can provide more sensitive areal delineation of aerosol events than can weather satellites. Shorter imaging distances (usually less than 400 km) and the use of oblique view angles to exaggerate the visual effect of aerosol loadings are the main reasons for this advantage.

Opportunities exist for both retrospective and prospective analyses using astronaut hand-held photographs of anthropogenic haze as an adjunct to surface and airborne observations. In situ observations characterize and quantify local forcing, while the photographs provide data on the geographic extent, temporal variability, and relationship to the existing meteorological conditions. Photographs often also indicate clearly the source regions for haze masses.

6.1.3 Significance of Desert Dust

Windborne dust in the atmosphere has also attracted increasing scientific interest in the last two decades. Until recently these interests were focused on local negative effects such as soil erosion, abrasion damage to crops and vehicles, negative impacts on the health of humans and livestock, and highway accidents (e.g., review in Pye, 1987). At a broader scale, geologists have long been interested in dust as a supplier of material for the rock type known as *loess*. Vast carpets of continent-derived dust particles are now known to blanket parts of the seafloor for distances of thousands of miles (Pye, 1987), providing information on the strength and direction of wind systems and even precipitation conditions during past glacial periods.

More recently, the rate and scale of regular dust transport has become better appreciated. Estimates suggest that 3000 metric tons (Mt) is lofted yearly into the atmosphere (Tegen and Fung, 1995; National Research Council, 1996) from numerous source regions (Figure 6.2). Approximately 20% of this reaches the oceans (e.g., Peterson and Junge, 1971), a volume similar to the sediment carried offshore annually by the Mississippi River (Ritter, 1968).

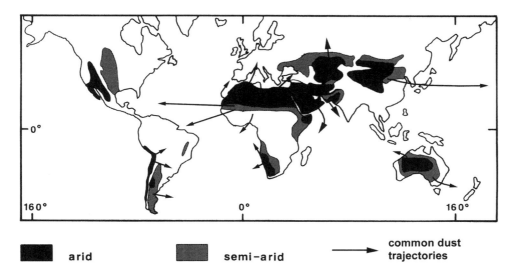

Figure 6.2 Global distribution of major deserts and surrounding semiarid regions, with some common dust transport trajectories. (Data from Coudé-Gaussen, 1984; Meigs, 1953; Wilkinson et al., 1998; F. Eckardt, R. Washington, M. J. Wilkinson, and K. P. Lulla, unpublished data.)

Due to the density of dust in the atmosphere during regular dust transport events, the forcing effect of dust on the energetics of the atmosphere—as with industrial pollutants—is beginning to receive attention. Heating and cooling effects depend on the loading, particle size, and land surface albedo (see the discussion in Idso, 1981). As a result, scientists developing global-change models have increasingly paid attention to atmospheric dust in attempts to quantify the effect of disturbed soils on climate (e.g., Tegen and Fung, 1994, 1995; Tegen et al., 1996). Interestingly, models suggest that roughly 50% of the dust loading in the atmosphere is derived from human disturbance of the soil surface and from vegetation-mat disruption due to short-term climate changes (Tegen et al., 1996).

Dust loadings in the atmosphere have been many times higher in the recent geological past, especially during drier and windier periods. Understanding modern dust source regions, transport mechanisms, and zones of deposition is critical to a better understanding of past dust behavior. The theory has even been advanced that dust blown from barren landscapes onto the major icecaps during the last glacial period might have reduced ice-surface albedo sufficiently to trigger long-term melting (Overpeck et al., 1996).

Another recent example of the key role dust is seen to play in global ecology is in biogeochemical cycling of materials between the continents. The best known example is the study of soil nutrients in the Amazon rain forest: Examining the limited amount of trace nutrients available to the rain forest ecosystem, Swap et al. (1992) have shown that dust fluxes from the Sahara Desert replenish soil nutrients in the Amazon lost to runoff.

6.2 EXAMPLES OF ANTHROPOGENIC HAZE PHENOMENA AND EVENTS

We present 21 photographs of 11 different industrial and dust aerosol events in various parts of the world (Table 6.1). The photographs were acquired on board the Space Shuttle and *Mir* space station. We illustrate one example of the integration of data from the photographs with other data, in the form of a case study of an air pollution event in the eastern United States [Figures 6.3 (see the color insert) and 6.4 to 6.7]. In this study various ground and in situ data demonstrate that the photographed haze in the atmosphere is undoubtedly air pollution rather than fog or biogenic haze. Linking photographs to related data illustrates the value of the photographs for meteorologic interpretation and spatial integration of ground data.

As the necessary supportive weather data were not available for conclusive interpretations, the other five examples (Table 6.1) rely on our interpretation that hazes are substantially due to pollution and its interaction with water vapor. However, the events we illustrate are very likely to have been air pollution, based on the visual appearance of the hazes, their location (which is usually a good indicator of the origin of the aerosol), and the broad weather patterns on the day of the event. We provide only one example of air pollution from the southern hemisphere. The southern hemisphere experiences only about 10% as much anthropogenic aerosol production as the northern hemisphere (National Research Council, 1996). Dust clouds are phenomena that are less controversial to interpret and are illustrated in the last four examples.

TABLE 6.1 Examples of Anthropogenic and Dust Aerosol Events

Feature	Figures	NASA Photographs
Industrial haze		
Eastern United States, April 1990	6.3 to 6.7	STS031-151-155
Northern Italy, August 18–26, 1997	6.8	NASA6-704-83, NASA6-707-65
Western Ukraine, October 19, 1997	6.9	NASA6-703-50
Red Basin, Sichuan, China, March 1996	6.10	STS075-721-22, STS075-773-66
Nile River delta, August 5, 1997	6.11	NASA5-707-52 and 53, STS084-310-36(inset)
Southern Africa, July 1997	6.12 and 6.13	NASA5-707-1
Dust		
Saharan dust over the western Mediterranean Sea, August 8, 1997	6.14 and 6.15	NASA5-708-48, 60, and 66
Tibetan Plateau, February 1997	6.16	NM22-759-329
Andean dust plumes, August 1983	6.17	STS008-46-936
Chinese dust in Oregon, April 1998	6.18 to 6.20	STS090-730-1, STS090-710-58, and STS090-739-72

6.2.1 Air Pollution and the Bermuda High, Eastern United States: Integrated Analysis of Photographic, Meteorologic, and Chemical Data

In this example we illustrate the types of photographs that are particularly useful for climate studies and the links that can be drawn with other data to provide greater understanding of pollution dynamics. The referenced photograph (Figure 6.3) was exposed on April 26, 1990. The Space Shuttle was located over Cuba at 27.5°N, 83.8°W at an altitude of 619 km (for orientation, see the approximate field of view superimposed on Figure 6.4A). For interpreting the photo, ancillary meteorological data are used to confirm (1) that conditions were appropriate for trapping haze, (2) that the observed aerosols are not clouds or fog, and (3) that the atmospheric chemistry was consistent with air pollution.

The stagnant high-pressure system (Bermuda High) located over Georgia and Florida (Figure 6.4A) provided an ideal meteorological situation for the accumulation of industrial haze in the lower levels of the atmosphere. The map of relative humidity reveals that the air was humid but *not* saturated at the surface or in the lower atmosphere, except over Florida (Figure 6.4B). Nor was extensive cloud formation possible at higher altitudes: Vertical temperature and dew point profiles (T and T_D), show that dew points at nine stations across eastern North America were below or far below ambient temperatures ($T_D < T$, or even $T_D \ll T$) (Figure 6.5). These profiles indicate that condensation could not occur and that only isolated altostratus clouds had formed. Despite the lack of fog or cloud, visibility was reduced significantly between April 25 and 27, 1990 (Figure 6.6) across the entire eastern United States.

Atmospheric chemistry is another parameter that can be employed to characterize the lower troposphere during the haze event of April 26, 1990 (Figure 6.3). "The

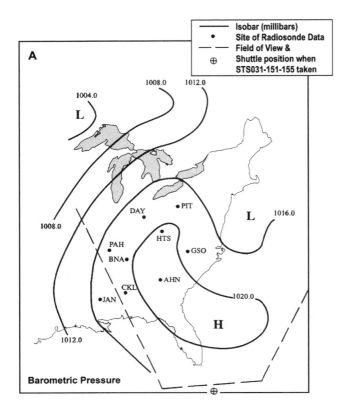

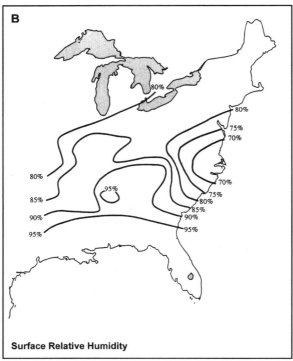

Figure 6.4 Meteorological conditions accompanying the haze shown in Figure 6.3 (color insert). (*A*) Barometric pressure at 12:00 GMT, April 26, 1990 (millibars). This pattern typifies the common summer pressure distribution known as the Bermuda High, with winds blowing clockwise around the center (H). Such a pattern would facilitate accumulation of atmospheric pollutants. AHN, Athens, Georgia; BNA, Nashville, Tennessee; CKL, Centreville, Alabama; DAY, Dayton, Ohio; GSO, Greensboro, North Carolina; HTS, Huntington, West Virginia; JAN, Jackson, Mississippi; PAH, Paducah, Kentucky; PIT, Pittsburgh, Pennsylvania (*B*) Surface relative humidity at 12:00 GMT on April 26, 1990. Although relative humidity is comparatively high throughout the eastern United States, fog development was possible only in the southernmost parts, excluding the possibility that the aerosols in Figure 6.3 were due to fog. (From the National Climatic Data Center, 1998b.)

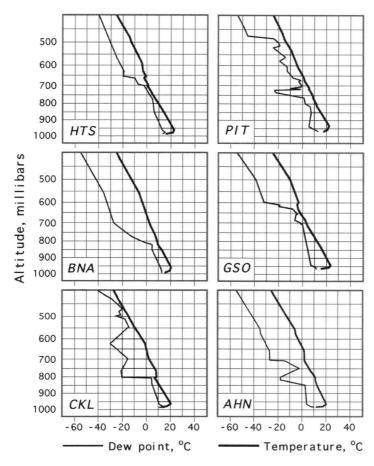

Figure 6.5 Vertical soundings of temperature (T, °C) and dew point (T_D, °C) versus altitude (mb) at 12:00 GMT on April 26, 1990 for six of the nine ground stations shown on Figure 6.4A. All soundings show $T_D < T$ such that relative humidity is less than 100% except in the extreme south, where humidity approaches 100% near the ground at Centreville, Alabama (CKL) and Jackson, Missouri (not shown). See city abbreviations in legend to Figure 6.4A. (From the National Climatic Data Center, 1998b.)

highly visible haze that persists in all of the industrial regions of the world consists mainly of sulfate and organic compounds from emissions of sulfur dioxide, organic gases, and smoke from biomass combustion" (National Research Council, 1996:9). Anthropogenic sulfur emissions are now estimated to account for 80 to 90% of the total sulfur load in the northern hemisphere, of which most is derived from coal burning and smelting of metal ores (National Research Council, 1996). Chemical data exist for periods of 24 hours or more before and after exposure of the photograph (Figure 6.3), but only a few sites were actively engaged in sampling at the time. Further, the chemical analyses were incomplete and did not account for all the aerosol mass. However, highly elevated levels of sulfate (SO_4^{2-}) over Bermuda (Figure 6.7), 1500 km east of Cape Hatteras, indicate that sulfate was indeed a major contributor to the reduced visibility values and to the increased albedo observed in

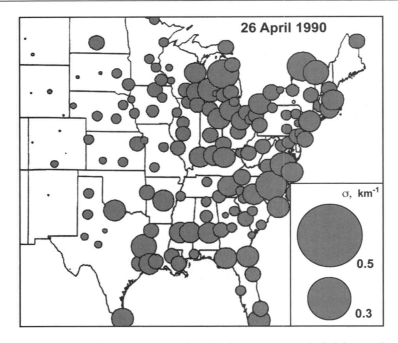

Figure 6.6 Visibility measurements show that the entire eastern third of the United States is under the influence of a widespread aerosol on April 26, 1990. Visibility is usually dominated by scattering of light and is estimated from visual range at numerous airports, shown here as daytime averages. Larger circles represent lower visibility (visibility is represented by the extinction coefficient, σ, which is proportional to the inverse visual range, in units of km^{-1}). (From the National Climatic Data Center, 1998b.)

the photo. Another likely component of the haze was organic matter from biomass combustion, due to burning of agricultural wastes and possibly to domestic fires.

We conclude therefore that the haze mass shown in the photograph (Figure 6.3) is indeed industrial haze. The haze is easily visible in the photograph because of the substantially increased albedo (reflectivity for sunlight). The photograph shows that the elevated sulfate measurements far out in the Atlantic were derived from regions directly upwind, namely in the industrial northeastern United States. The high-pressure system of clockwise winds not only transported the haze into the Atlantic, but then returned it toward the Florida peninsula. The leading edge of the haze mass can be seen on the far right of the view, about to affect air quality in the Bahamas and then in the metropolitan region of southern Florida (Figure 6.3).

The fact that the aerosol mass remained as a coherent, recognizable corridor despite the great distance some of the particles had apparently been transported was surprising to astronaut crews and researchers alike.

6.2.2 Other Probable Pollution Hazes

Northern Italy, October 18–26, 1997. Some of the most dramatic photographs of the densest and most permanent industrial pollution on the planet come from the

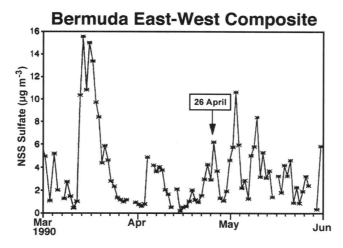

Figure 6.7 Sulfate content of atmospheric aerosol at Bermuda for March 1 to June 1, 1990. Note that even higher loadings were measured at Bermuda on three other occasions within the few weeks before and after the event described here. (Data from J. Prospero, University of Miami, Miami, Florida, 1999, personal communication.)

Po River valley. The valley floor is often invisible for days from low Earth orbit, while the flanks and peaks of the Alps are very clear. Crews have often documented a well-defined high-albedo "river" of thick polluted fog flowing out of the mouth of the Po valley, which is then channeled down the entire length of the Adriatic Sea.

Photographs taken during the NASA 6 mission documented a week-long pollution event related to the Po valley industrial region, starting on October 18, 1997 and continuing until October 26 (Figure 6.8; see the color insert). Ground measurements show that visibilities during this event dropped as low as 2.8 km at Venice on October 21 (Anonymous, 1998b). Later in the event, a visual discontinuity appeared between zones of different haze loadings (arrows, Figure 6.8). Although the line is clear in these photographs, surface visibility readings barely detected the difference: Readings were within the noise level for such visibility measures, which were 10.5 km (6.5 miles) ahead of the aerosol front and 9.6 km (6.0 miles) behind it (Anonymous, 1998b).

Western Ukraine, October 19, 1997. Barometric data indicate that in mid-October, 1997, 850-mb gradients were nearly nonexistent, with a flat high pressure ridge from northern Italy to Kiev (Anonymous, 1998a,b). These conditions served to contain pollutants near the ground in sufficient concentration to be documented on film.

The flow in the boundary layer in the foreground was northwesterly, that is, from lower right to upper left in Fig. 6.9, parallel with the air-pollution tendrils. The tendrils remain coherent over great distances possibly because winds were light. Tendrils were aligned with the Carpathian mountain front (C–C, Figure 6.9), suggesting localized topographic channeling of the air mass by the mountain range. This degree of channeling is also affected by the high-pressure system since low-pressure conditions would serve to mix the lower atmosphere more thoroughly, as well as to allow airflow to override some topographic irregularities. Higher concentrations of aerosols (H) accumulated in the area to the east of the Carpathians in the lower Danube River basin of southern Romania, a topographic basin in which haze from the Ukraine appears to have accumulated on this day.

Upper-air circulation at 500 mb showed a ridge of high pressure extending from the Carpathian Mountains northeast to 52°N 50°E (Anonymous, 1998a,b). This ridge, with its associated weak pressure gradients, agrees with the apparent atmospheric stagnation in the southern half of the photograph, evidenced by denser aerosol loadings in all topographic basins.

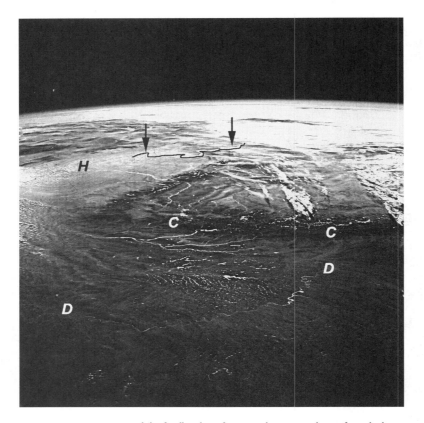

Figure 6.9 Numerous tendrils of polluted air (foreground) move southeast (from the lower right), from the western Ukraine toward Romania, maintaining their coherence for tens of kilometers on October 19, 1997 (13:02:42 GMT). Anthropogenic haze (H) also occupies the lower Danube basin in southern Romania, a topographic basin in which the Ukrainian haze appears to be accumulating. The Carpathian Mountains (C–C) appear with greater clarity above the inversion layer (below which lies the polluted air). The sun is reflected off rivers, revealing details of the Dniester River meanders (D–D), which represents a distance of about 280 km. The distance from the Dniester to the Danube River at the arrows is about 600 km (arrows indicate the bend of the river at the Iron Gate on the Yugoslav border). (NASA photograph NASA6-703-50, center point 48.5°N 25.5°E, craft nadir 51.2°N 34.7°E, Hasselblad camera, 100-mm lens, altitude 383 km.)

Red Basin, Sichuan, China, March 1996. The influence of topographic concentration of air pollution is well illustrated by air pollution events in China. Under anticyclonic conditions—such as those developing when Figure 6.10A was taken—the floor of the Red Basin is often obscured under a blanket of gray haze, whereas surrounding mountains appear in great detail under clear air (Figure 6.10A; see the color insert). Sichuan Province, China's largest, is home to more than 100 million people and two rapidly industrializing regions.

An unexpected benefit of the attempt to photograph this pollution episode during the STS-75 mission (late February to early March 1996) led to the unique view of a wide (about 200 km) but coherent plume of anthropogenic haze rising from the eastern provinces of China. The haze ascends into the upper westerly winds and extends at least 600 km into the Pacific Ocean (Figure 6.10B). Such plumes are probably a common feature of the atmosphere downwind of the rapidly industrializing regions of eastern China.

Nile River Delta, August 5, 1997. Rapidly growing megacities in the tropics act as vast, visible point sources in the developing world for air pollution masses hundreds of miles long. Oblique views show these features well in the vicinity of northwestern Taiwan, Rio de Janeiro, Managua, Cairo, Bangkok, and Surabaya in Indonesia, all of which commonly show long corridors of pollution downwind (M. J. Wilkinson, unpublished data).

One of several examples of industrial air pollution from the Shuttle–*Mir* photographs is pollution over the Nile delta in two overlapping views (Figure 6.11; see the color insert). Almost 30 million people live in the restricted area of the Nile delta, which measures 160 km from its base where Cairo is located, to the Mediterranean Sea (top left, Figure 6.11). The dark green irrigated agriculture indicates the edge of the delta and Nile River.

Cairo lies unseen under the densest part of the haze at the apex of the delta (Figure 6.11). By measures of suspended particulates and lead pollution, Cairo is the most heavily polluted large city in the world (U.S. Agency for International Development, 1998), with emissions from factories, construction projects, garbage burning, automobiles, and desert dust as the major contributors. An estimated 10,000 to 25,000 deaths each year are attributed to air pollution, and studies show adverse health effects on children who live in Cairo (U.S. Agency for International Development, 1998). The size and density of the rural population undoubtedly also contribute to the haze in the form of dust and biomass burning.

The continuous layer of smoke and haze covering the delta obscures almost all detail of the road and city patterns. By contrast, details can be seen easily on the brown surface of the desert, especially upwind on the west side of the delta (Figure 6.11). This observation leaves almost no doubt that the loss of detail over the delta in these pictures is due to anthropogenic materials ejected into the atmosphere. Under different weather conditions, the anthropogenic hazes are blown away, giving clear views of the city (Figure 6.11, inset).

The reason for the sharp boundary between clear and polluted air is apparently due to the local topography—the polluted air is retained within the delta basin (only 100 m below the desert surface near Cairo). We surmise that an atmospheric inversion at roughly the altitude of the desert floor was concentrating pollutants within this thin layer of air. Other examples of pollution masses related to tropical cities (not shown here) include Taiwan (NM22-716-24), Bangkok in southern Thailand (NM22-778-103), Santiago, Chile (NM23-720-640), and the Brazilian coast at São Paulo (NASA6-708-79).

Southern Africa, July 1997. Southern Africa might be the most vigorous supplier (highest "source strength") of aerosols to the atmosphere in the southern hemisphere. Research has already shown that 135 Mt/yr of aerosol is delivered into the atmosphere over southern Africa (Garstang et al., 1996). Aerosols and trace gases (CO, NO_x, CH_4BR) in unexpectedly high concentrations are generated as natural biogenic haze, as smoke from anthropogenic biomass burning, and from urban-industrial air pollution. Pollution has reduced visibility in the Johannesburg region to as little as 250 m (R. Swap, University of Virginia, Charlottesville, 1996, personal communication). Southern Africa is probably the major contributor to the largest CO_2 and low-level ozone anomaly in the southern hemisphere, namely the anomaly in the tropical South Atlantic (Fishman, 1992).

Research in air-mass evolution in South Africa (Garstang et al., 1996) has recently described aerosol sources, air-mass residence time, and trajectories by which hazes exit the subcontinent. Some weather situations transport hazes offshore fairly directly. Other weather situations under which the hazes become particularly dense result from recirculating wind systems which maintain the hazes onshore under clear skies (Figure 6.12) so that chemical evolution can proceed (Garstang et al., 1996). Air masses can recirculate onshore several times before being transported off the subcontinent, typically via the southeast Africa "exit" to the southern Indian Ocean (Figure 6.12), but also into the Atlantic Ocean (Garstang et al., 1996).

The semistationary high-pressure systems over the subcontinent provide both the capping mechanism to hold aerosols and pollutants near the ground, thereby increasing concentrations, as well as providing the continuous circulation under which chemical evolution of the pollutants occurs. Evacuation of pollution haze via the southeastern route is well illustrated in Figure 6.13, although the direction of movement is more southerly than the norm indicated in Figure 6.12.

Researchers connected to the NASA-sponsored series of experiments known as SAFARI 92 (Andreae et al., 1996) and SAFARI 2000 continue to investigate from ground-based, airborne, and satellite observations the source regions in South America and southern Africa of aerosols and trace gases, and in turn, their effects on cloud formation and regional climate through atmospheric forcing by these substances (R. Swap, University of Virginia, Charlottesville, 1998, personal communication).

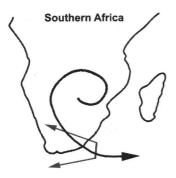

Figure 6.12 Anticyclonic circulation of the atmosphere over southern Africa typically evacuates pollutants offshore to the southeast (another evacuation route lies over Angola). Footprint of the oblique, west-looking photograph shown in Figure 6.13 is indicated. Note that cyclonic/anticyclonic motions in the southern hemisphere display opposite rotations to those in the northern hemisphere. (Modified from Garstang et al., 1996.)

Figure 6.13 Highly discrete corridor of aerosols streaming off the southeast coast of South Africa (foreground) on July 31, 1997, 13:48:14 GMT. The southern coast of Africa occupies the middle of this west-looking view, with Cape Point (near Cape Town) in the distance at the angle of the coast. (NASA photograph NASA5-707-01, center point 33°S 23°E, craft nadir 33.1°S 35.7°E, Hasselblad camera, 250-mm lens, altitude 391 km.)

6.3 EXAMPLES OF DUST PHENOMENA AND EVENTS

Saharan Dust over the Western Mediterranean Sea, August 8, 1997. A large and dense plume of dust obscures both land and water in the western Mediterranean basin in a series of photographs (NASA5-708-48, -52, -55, -57, -60, -66) taken in early August 1997, three of which are shown here in views with increasing levels of detail and more vertical look angles (note that more ground detail is increasingly visible as the look angle approaches the vertical, Figure 6.14; see the color insert). Some of the dense dust is apparently related to clusters of storm cells. This is not unusual except for the fact that the storm cells are situated over the Mediterranean Sea. Linear structures in the dust radiate from the cloud mass. Gravity waves, oriented orthogonally to the linear structures, appear at various points, especially related to the Balearic islands [view (C)]. Smaller cloud masses are also associated with the islands [I in view (A), P in view (B)].

A tongue of high pressure extended from north Africa (Tunisia) to Sicily on August 6, 1997. Consequently, winds at 700 and 850 mb over Algeria continued to be southwesterly, strengthening to 15 to 20 knots (28 to 37 km/h) by August 8, 1997 (Anonymous, 1998c), transporting the dust from the Sahara. Visibility was much reduced in Algerian coastal cities such as Annaba, Bejala, and Tiaret as early as August 7, 1997, lasting through August 10. Most stations in the western Mediterranean also showed this event was the most intense of the month of August (Figure 6.15). Further east, visibility at stations in Tunisia—Kairouan and Bizerta—was not reduced, suggesting that these stations were not influenced by the dust storm, and indicating that the plume was about 800 km wide. Visibility for the town of Biskra on the *south* flank of the Atlas mountains was reduced as early as August 7 (Figure 6.15), indicating both the Saharan source of the dust mass and the fact that this major event lasted more than 24 hours. Stations in Sardinia (Olbia and Cagliari) showed visibility minimums on August 10, 1997. This delay is consistent with dust transport via gentle west-southwesterly flow (10 to 15 knots; 18.5 to 28 km/h) at 850 mb (Anonymous, 1998c).

Tibetan Plateau, February 1997. As part of the Shuttle–*Mir* program, aerosol events were photographed (1) in locations where they have not been documented in

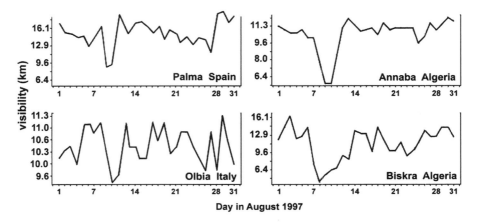

Figure 6.15 Visibility plots (miles) for three cities in the Mediterranean basin (Annaba, Algeria; Palma, Spain; Olbia, Italy) and one trans-Atlas city (Biskra, Algeria) for August, 1997. (From the National Climatic Data Center, 1998a.)

hand-held photographs before, (2) on a scale that had not been appreciated, or (3) that were too small to be identified on weather-satellite images, especially over land surfaces. A good example is the documentation of major but rare dust clouds embedded in westerly winds on the Tibetan Plateau. The event lasted two days in February 1997 (Figure 6.16, see the color insert).

Dust and snow accumulation are the two critical components in the ice layers of icecaps in the tropics which yield evidence of past climatic change. Icecaps in Tibet are being drilled for paleoclimatic information related to annual layers of snow and dust. This research suggests, surprisingly, that climatic histories in the last 1500 years have been similar in the highlands of Tibet and the high tropical Andes in South America (Thompson, 1996).

Andean Dust Plumes, 1982 El Niño. Based on astronaut photographs, dust events have been documented streaming off the high, arid Altiplano–Puna plains of the central Andes Mountains in northwest Argentina and Bolivia since 1973. Hand-held photographs have provided the basic data (25 events) for the description of this dust-flux system: Dust is transported by westerly winds every southern winter. Wind-vector anomalies at 500 mb indicate that dust plumes occur, as expected, under winds of above-average strength (M. J. Wilkinson, R. Washington, and K. P. Lulla, unpublished data). A frame exposed during the NASA 2/*Mir* 21 flight in June 1996 (NM21-760-39, not shown) provided another data point. Figure 6.17 (see the color insert), however, illustrates an unusually large set of dust plumes, the size of which might be related to atmospheric anomalies induced by the extreme 1982–1983 ENSO event (Clapperton, 1993).

The hand-held photographs provide new perspectives on the movement of dust and its influence on physical environments in South America (Wilkinson et al., 1998). (1) Some dust is driven from the Atacama Desert in Chile across the high Andes, to fall in the Chaco plains on the east side of the Andean massif. (2) Some plume trajectories are oriented toward the southern Amazon basin, which suggests that Africa is thus not the only source of trace minerals transported into the Amazon basin, as presently hypothesized (Swap et al., 1992). (3) Iriondo (1993) believed that loess beds occupying large areas in lowland northern Argentina and southern Bolivia were derived from source areas to the north or south, whereas our photographic data indicate that the loess is derived at least partly from the Altiplano.

In Figure 6.17, the main eastward-blowing plume intersects the dark Cachi massif in the middle of the view at about 6000 m. Dust transported at this altitude in strong winds is typically carried for many hundreds and often some thousands of kilometers (Bergametti, 1997). It is therefore possible for Altiplano dust to reach the Antarctic since the same westerly wind system in Patagonia is known to provide dust to Antarctica (Basile et al., 1997).

Chinese Dust in Oregon, June 1996. Winds in western China (Xinjiang Province) raised a major dust storm on April 17, 1998. The dust was carried thousands of kilometers to the western states of North America, where it arrived some 10 days later (Figure 6.18). Such events are sufficiently unusual that an ad hoc committee set up a Web site to describe the event (Husar, 1998). Hand-held astronaut photographs documented the event in Asia (Figure 6.19A), in its passage over the open ocean (Figure 6.19B), and as a haze in California and Oregon (Figure 6.19C). The complete series of views from low Earth orbit (STS090-702-51, STS090-702-87, STS090-710-58, STS090-730-1, STS090-735-65) uniquely complements satellite images of the event (Figure 6.20), at every point from the Gobi Desert, over the Sea of Japan and the Japanese archipelago, over the open ocean, and then to the western United States. Figure 6.20 shows a wider view of this very large event from the GOES 10 satellite.

Degraded air quality was noted along the entire west coast of North America, starting in Vancouver, Canada, and extending 1600 km in subsequent days to southern California. Although first attributed to prescribed burning in the states of Idaho, Washington, and Oregon (which in turn led to the banning of all set fires in some communities), it was later announced that at least part of the pervasive haze could be ascribed to fine dust transported from China (Husar, 1998).

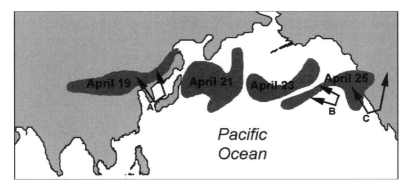

Figure 6.18 Trans-Pacific dust movement from the Gobi Desert to North America. Positions of the first in a series of dust emissions from Asia are indicated by dates between April 19 and 25, 1998. Arrows indicate locations of photographs in Figure 6.19. The data were acquired from SeaWiFs, GMS, and three GOES satellites. (Modified from Husar, 1998.)

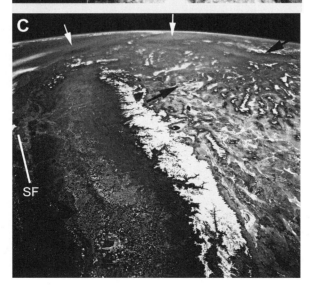

Figure 6.19 Movement of dust from the Gobi Desert across the pacific to North America. (*A*) Margin of the dust plume on April 22, 1998 (03:12:52 GMT), 4 days after the leading edge of the Gobi dust first exited Asia (arrows). The Korean Peninsula appears under cloud down the left side of this north-looking view. The coast of North Korea stretches across the top (dashed line). (*B*) Gobi Desert dust over the northeast Pacific Ocean photographed about 2500 km north-northeast of Hawaii, on April 25, 1998 (20:59:22 GMT). Here dust was transported east (from the center of the horizon toward the lower left) over the Pacific Ocean in association with the jet stream (dense line of cloud). By this date the dust had been transported nearly 5000 km from its source in central Asia. (*C*) Gobi Desert dust formed a prominent haze over the Pacific Northwest on April 27, 1998 (20:58:31 GMT). The densest dust is indicated in this figure (and Figure 6.20) by white arrows—the left arrow shows a concentration offshore and the upper center arrow shows a concentration over the Columbia Basin of Washington state. The black arrows in this view (and arrows in Figure 6.20) indicate the position of a regional dust front, north of which air over the United States and southern Canada was distinctly hazy. This view was taken from a point over Los Angeles and shows the central valley of California in the foreground. [NASA photographs, Hasselblad camera, 40-mm lens: (*A*) STS090-730-01, center point 40.5°N 128.5°E, craft nadir 35.3°N 132°E, altitude 251 km; (*B*) STS090-710-58, craft nadir 38.2°N 48.8°W, altitude 246 km; (*C*) STS090-739-72, center point 38°N 119.5°W, craft nadir 35.2°N 119.1°W, altitude 244 km.]

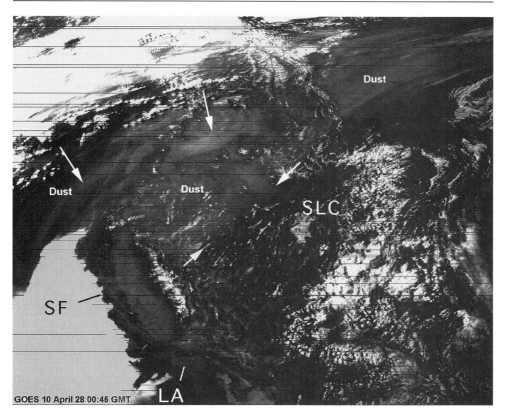

Figure 6.20 This GOES 10 satellite visible image was acquired less than 4 hours after the view shown in Figure 6.19C. For features marked by arrows, see caption to Figure 6.19C. The dust mass extended well into the High Plains states and south-central Canada (top right). Dark horizontal lines are due to data loss. (Modified from Husar, 1998.)

6.4 CONCLUSION

Dust and anthropogenic hazes are sufficiently important in understanding the energetics and transport processes of the atmosphere that photographs of haze features tens to thousands of kilometers in diameter hold particular interest. For research scientists, photographs taken by astronauts stimulate three-dimensional visualization of atmospheric phenomena. These photographs also allow point-source data to be integrated over wide geographic areas. Industrial hazes are sufficiently difficult to define geographically from automated satellite images that the oblique views, especially in photographs taken by astronauts, have special significance for this research. They also reveal subtle internal features within haze masses, which can be important to understanding the development of some events. These images are at scales that show components of larger systems that are difficult to image either from the ground or from weather satellites.

Astronaut photographs taken from orbiting spacecraft also permit atmospheric discoveries—data showing the existence of phenomena in unexpected places, or of unexpected intensity, size, or behavior. Crews using hand-held cameras have documented the existence, precise location, and dimensions of aerosol events unsuspected from weather satellite data. Human beings who are in orbit continuously for months at a time have the advantage of being able to document some larger, dynamic features of the atmosphere where time scales vary from hours to years.

We illustrated haze events that lasted from a few days to two weeks in various parts of the world. After a case study of a haze event in the eastern United States that integrated meteorological data with the aerosol event shown in a photograph from low Earth orbit, five examples of probable anthropogenic hazes were presented, some with associated ground data. We presented four dust events from three continents. The Tibetan and Andean examples showed regional events, with dust sources in arid regions affecting the atmosphere and surface geology up to hundreds of kilometers downwind. In the Saharan and Chinese examples, photographs of parts of very large events were presented in which a nearby continent (Europe) and a distant continent (North America) acted as dust sinks.

Our experience shows that students, researchers, and the general public alike find astronaut photography of Earth a valuable source of information. Previous difficulties in examining the photographs remotely—and at resolutions that reveal more of the detail on the master film—are receding as technology promotes the more rapid transmission of large amounts of digital data. We anticipate more widespread use of these data in the near future.

ACKNOWLEDGMENTS

We acknowledge the many astronauts and cosmonauts for their enthusiasm, care, and expertise in taking these and many other photographs of Earth. We thank reviewers for the improvements they suggested, especially P. W. Dickerson and C. A. Evans for this and much other assistance. A. Spraggins provided data on some of the frames.

REFERENCES

Andreae, M. O., Fishman, J., and Lindesay, J. 1996. The Southern Tropical Atlantic Region Experiment (STARE): Transport and Atmospheric Chemistry near the Equator-Atlantic (TRACE A) and Southern African Fire-Atmosphere Research Initiative (SAFARI): an introduction. *Journal of Geophysical Research*, 101(D19):23519–23520.

Anonymous. 1998a. Weather data for 19.10.1997 [original in German], *Deutsche Wetterdienst* (European Meteorological Bulletin, Offenbach, Germany), 22(292):1–7.

Anonymous. 1998b. Weather data for 21.10.1997–27.10.1997 [original in German], *Deutsche Wetterdienst* (European Meteorological Bulletin, Offenbach, Germany), 22(294–300):1–7 in each issue.

Anonymous. 1998c. Weather data for 6.8.1997–9.8.1997 [original in German], *Deutsche Wetterdienst* (European Meteorological Bulletin, Offenbach, Germany), 22(218–221):1–7 in each issue.

Basile, I., Grousset, F. E., Revel, M., Petit, J. R. Biscaye, P. E., and Barkov, N. I. 1997. Patagonian origin of glacial dust deposited in East Antarctica (Vostok and Dome C) during glacial stages 2, 4 and 6. *Earth and Planetary Science Letters,* 146:573–589.

Bergametti, G. 1992. Atmospheric cycle of desert dust, in W. A. Nierenberg (ed.), *Encyclopedia of Earth System Science,* 1:171–182. New York: Academic Press.

Charlson, R. J., and Wigley, T. M. L. 1994. Sulfate aerosol and climatic change. *Scientific American,* 270(2):48–57.

Charlson, R. J., Langner, J., Rodhe, H., Leovy, C. B., and Warren, S. G. 1991. Perturbation of the northern hemisphere radiative balance by backscattering from anthropogenic sulfate aerosols. *Tellus,* 43AB:152–163.

Clapperton, C. 1993. *Quaternary Geology and Geomorphology of South America.* New York: Elsevier, 779 pp.

Coudé-Gaussen, G. 1984. Le cycle des pouissières éolienne désertique actualles et la sédimentation des loess peridésertiques quarternaires. *Bulletin du Centres de Récherches sur les Exploration-Production ELF-Aquitaine,* 8:167–182.

Fishman, J. 1992. Probing planetary pollution from space. *Environmental Science and Technology* 25:613–621.

Garstang, M., Tyson, P. D., Swap, R., Edwards, M., Kållberg, P., and Lindesay, J. A. 1996. Horizontal and vertical transport of air over southern Africa. *Journal of Geophysical Research,* 101(D19):23721–23736.

Hall, R., 1998. Using 3D orthophotos to increase communication at public presentations. Natural Resource Management Using Remote Sensing and GIS, *Proceedings, 7th Forest Service Remote Sensing Applications Conference,* Houston TX, April 6–9, pp. 390–391.

Husar, R. 1998. Windblown dust from Asia hits North America: the Asian dust event of April 1998.
http://capita.wustl.edu/Asia-FarEast (viewed December 16, 1998)

Kaspar, P. 1998. High-performance networks provide access to NASA's data treasure-trove. *Insights* (NASA, Moffett Field California), 5:12–17.

Kiehl, J. T., and Briegleb, B. P. 1993. The relative roles of sulfate aerosols and greenhouse gases in climate forcing. *Science,* 260:311–314.

Idso, S. B. 1981. Climatic change: the role of atmospheric dust. In Pewe, T. L., ed., *Desert Dust: Origin, Characteristics, and Effect on Man.* Special Paper 186. Boulder, CO: Geological Society of America, pp. 207–216.

Iriondo, M. 1993. Geomorphology and late Quaternary of the Chaco (South America). *Geomorphology,* 7:289–303.

Meigs, P. 1953. World distribution of arid and semi-arid homoclimates. *Proceedings, Ankara Symposium on Arid Zone Hydrology, UNESCO,* Paris, vol. 1, pp. 203–209.

Mudur, G. 1995. Monsoon shrinks with aerosol models. *Science,* 270:1922.

National Climatic Data Center. 1998a. Online climate data.
http://www.ncdc.noaa.gov/ol/climate/climatedata.html (viewed February 26, 1998)

National Climatic Data Center. 1998b. Regional climate centers [active archive of weather data]. *http://www.ncdc.noaa.gov/ol/climate/aasc.html#RCC* (viewed December 1, 1998)

National Research Council. 1996. *Aerosol Radiative Forcing and Climate Change: A Plan for a Research Program.* Washington, DC: National Academy Press, 161 pp.

Overpeck, J., Rind, D., Lacis, A., and Healy, R. 1996. Possible role of dust-induced regional warming in abrupt climate change during the last glacial period. *Nature,* 384:447–449.

Peterson, J. T., and Junge, C. E. 1971. Source of particulate matter in the atmosphere. In Matthews, W. H., Kellogg, W. W., and Robinson, G. D., eds., *Man's Impact on Climate.* Cambridge, MA: MIT Press, pp. 310–320.

Pye, K. 1987. *Aeolian Dust and Dust Deposits.* New York: Academic Press, 334 pp.

Ritter, D. 1968. Continental erosion. In Fairbridge, R. W., ed., *Encyclopedia of Geomorphology.* Stroudsburg, PA: Dowden, Hutchinson & Ross, pp. 169–174.

Swap, R., Garstang, M., Greco, S., Talbot, R., and Kållberg, P. 1992. Saharan dust in the Amazon basin. *Tellus,* 44B:133–149.

Tegen, I., and Fung, I. 1994. Modeling of mineral dust in the atmosphere: sources, transport, and optical thickness. *Journal of Geophysical Research,* 100(D9): 22897–22914.

Tegen, I., and Fung, I. 1995. Contribution to the atmospheric mineral aerosol load from land surface modification. *Journal of Geophysical Research,* 100(D9): 18707–18726.

Tegen, I., Lacis, A., and Fung, I. 1996. The influence on climate forcing of mineral aerosols from disturbed soils. *Nature,* 380:419–422.

Thompson, L. G. 1996. Climatic changes for the last 2000 years inferred from ice-core evidence in tropical ice cores. In Jones, P. D., Bradley, R. S., and Jouzel, J., eds., *Climatic Variations and Forcing Mechanisms of the Last 2000 Years.* NATO ASI Series, Vol. 141. Berlin: Springer-Verlag, pp. 281–295.

U.S. Agency for International Development. 1998. USAID/EGYPT: protecting the environment overview. *http://www.usaid-eg.org/env-ovr.htm* (viewed November 28, 1998)

Wilkinson, M. J., Gonzalez Bonorino, G., German Viramonte, J., and Lulla, K. P. 1998. Space Shuttle photography documents seasonal dust flux from the Andes Mountains (15–28°S). *Abstract, Loess in Argentina: Temperate and Tropical, International Joint Field Meeting,* Paraná, Argentina, May 15–21.

Chapter 7

Biomass Burning and Smoke Palls with Observations from the Space Shuttle and Shuttle–*Mir* Missions

M. Justin Wilkinson
Office of Earth Sciences
NASA Johnson Space Center
Houston, Texas USA

Kamlesh P. Lulla
Office of Earth Sciences
NASA Johnson Space Center
Houston, Texas USA

Marvin Glasser
Department of Physics and Physical Sciences
University of Nebraska
Kearney, Nebraska USA

ABSTRACT

We surveyed photographs taken by cosmonauts and astronauts since 1992, including photographs taken from the *Mir* space station and Space Shuttle during the Shuttle–*Mir* program (March 1996, to June 1998), to determine the contribution of high-latitude long-duration space missions toward understanding the global distributions of biomass burning. The Shuttle–*Mir* photographs and associated data provided spatial and temporal information on biomass burning for comparison with previous assessments of this environmental phenomenon. Because of the high orbital inclination of Space Shuttle and *Mir* flight paths, and the more continuous temporal coverage possible from *Mir*, better data on biomass burning were obtained for high latitudes and other parts of the world that had not been well covered in previous studies. Conclusions broadly support earlier views of seasonality and dominant controls. Photos of major fires in Russia–Mongolia, Southeast Asia, Mexico–Central America, and south-central Africa during the period of the Shuttle–*Mir* program give a visual impression of the great extent of certain smoke palls. In Southeast Asia and Mexico, the quantity of smoke was related to the extreme El Niño event of 1997–1998. Spacecraft in low Earth orbit provide an intermediate platform for biomass-burning observations between ground observations/aircraft views and satellite images.

7.1 INTRODUCTION

The physical and ecological environments of Earth are changing vigorously and rapidly (Botkin et al., 1989). Of major scientific interest are questions concerning change in biological and atmospheric systems and the influence of anthropogenic modifications of the environment on these changes. Botkin et al. (1989) suggested that the extrapolated rate of present environmental change is more rapid than any known from the geological record, including the great waves of biological extinction recorded at some major geological-era boundaries.

Global biomass burning is an increasingly important agent of contemporary environmental change (Levine, 1990a,b). Prior to the nineteenth and twentieth centuries, anthropogenic biomass burning in savannas, dry–wet tropical forests, and Asian lowland dipterocarp forests was a traditional practice with relatively low biological impact, as well as being a natural event that occasionally reached regional proportions (Pyne, 1995).

Biomass burning contributes significantly to the atmospheric budget of particulates and trace gases. Andrasko et al. (1990) reported that global deforestation and biomass burning contribute as much as 15% of the current anthropogenic emissions of greenhouse gases. The amount of data being collected and published on tropical biomass burning is increasing (Booth, 1989; Andreae, 1990). Information is available on biomass burning in regions of traditional savanna management, and in tropical moist forests where conversion to agricultural and grazing uses takes place, as in South and Southeast Asia, Africa, and Latin America (Levine, 1991). However, the geographical distribution and temporal frequency of biomass burning in high-latitude forests are much less well known. Modeling atmospheric, hydrologic, and ecological changes connected with large-scale, repetitive biomass burning is hampered by a lack of basic geographic, temporal, chemical, and physical data.

Previous studies of biomass burning using Space Shuttle photographs have presented analyses of smoke-pall size, documenting the largest areas covered by regional smoke in South America (Helfert and Lulla, 1990) and Southeast Asia (Lulla and Helfert, 1989). One of the largest smoke palls ever recorded (Cachier, 1992) covered most of the Amazon basin in 1985 and measured at least 3,500,000 km^2 (Helfert and Lulla, 1990). The 1991 Amazon smoke pall, also documented on Shuttle photographs, covered more than 2,300,000 km^2 (Lulla et al., 1994). Helfert and Lulla (1990) documented the expansion in pall size through successive seasons in South America and showed that palls expanded from a few hundred square kilometers in area in the early 1980s to subcontinental size in the course of only one decade. Another research effort (Andreae, 1993) focused on extracting information on geographical and temporal patterns in biomass burning from the dataset of Space Shuttle photographs.

Fires and smoke were unusually prominent in the photographs acquired during the Shuttle–*Mir* program (March 1996 through June 1998). Biomass burning was widespread process at various times during the program, especially during months when burning was exacerbated by severe droughts resulting from the 1997–1998 El Niño–Southern Oscillation (ENSO; e.g., *International Forest Fire News,* 1998). Our main objective is to illustrate the potential of the growing body of Space Shuttle and *Mir* photographic data for temporal and spatial studies of global biomass burning. We build on Andreae's (1993) work by analyzing subsequent data acquired by astronauts. In addition to routine observations from missions unrelated to *Mir,* the Shuttle–*Mir* missions provided opportunities for astronauts to record global environmental changes such as biomass burning on a continuing basis.

7.2 METHODS

Andreae's (1993) study yielded 1030 images of biomass-related fire and smoke taken on missions between May 1962 (*Gemini 5*) and May 1992 (STS-49), culled from more than 100,000 photographs in the database maintained by what is now the Office of Earth Sciences (OES, 1999) at the NASA Johnson Space Center in Houston. Glasser and Lulla (1997) surveyed the same database through November 1995 (STS-74), by which time the total number of records had risen to 261,000. This study is based on 369,682 frames in the database (through STS-90, including all Shuttle–*Mir* missions), of which 3745 frames recorded biomass burning, an almost fourfold increase over the number of observations available to Andreae (1993) in May 1992.

The OES database holds cataloged data on each photograph and includes a unique identification number, date and time (GMT) of acquisition, latitude and longitude of both photograph center point and spacecraft position, and a brief description of the main feature in the photograph (other fields include lens focal length, angle of view, spacecraft orbit, and degree of cloud cover, among others; see Lulla et al., 1996).

For the sake of comparability, we broadly follow Andreae's (1993) methods for examining the spatial and temporal occurrence of biomass burning observed globally. We searched the OES photograph catalog for keywords based on the text strings "fire," "smoke," and "burn." From this subset we excluded oil, urban, and industrial fires, where these could be identified from the descriptions, or from locations over major urban areas as interpreted from the latitude/longitude field. As the great majority of fires was reported from areas outside the largest industrial and urban regions of the planet, we are confident that relatively few cases that were not biomass burning have been included in our raw data.

Catalogers use a number of features to identify smoke accurately in space photographs. Individual point sources of smoke (and even flame) can be seen, with characteristic plume expansion as the smoke rises through the atmosphere. In contrast to linear cloud formations, smoke plumes diverge downwind from the associated shadows cast on the ground, indicating that the plume originates on the ground surface. Blackened fire scars at the point source often confirm the interpretation. Furthermore, the texture and color of clouds usually differ from those of smoke, and this is obvious where clouds and smoke appear in the same frame.

For large smoke palls that might be confused with stratus clouds, we used news reports to confirm the existence of very large fires. Through our experience over the two decades of more intensive cataloging of smoke at different scales, and in understanding the seasonality of burning in different biomes, we believe that smoke has not been misinterpreted systematically or on a scale that would invalidate the analyses presented here, although no tests of cataloger error have been conducted. In earlier studies of smaller datasets, the interpretation of all fire/smoke photographs was rechecked from the film (Andreae, 1993; Glasser and Lulla, 1997). Although we were unable to perform a similar check of all the data since May 1992, we nevertheless believe that our broad conclusions have not been compromised.

Based on photographs taken during Space Shuttle and previous programs from 1961 through November 1995 (STS-74), Glasser and Lulla (1997) showed that relative frequency (the number of biomass burning photographs in any year divided by the number of photographs taken in that year) was a better measure of assessing biomass burning than was the absolute number of biomass burning photographs. The relative frequency averages 1.2% for OES data up to 1995, but it varies widely, from 10.2% for STS 51 to 0% for STS-54. Glasser and Lulla (1997) showed that this statistic was not dependent on the duration of a Shuttle flight, the number of photographs exposed, or the inclination of the flight path. Further, the latitudinal distribution of biomass burning in the photographs was in accord with characteristic vegetation distribution and distributions of biomass burning as far as these data are known.

In this preliminary study we have not performed any normalizing statistical analysis. We merely aim to show which major gaps in the data have been addressed by Shuttle–*Mir* and prior Space Shuttle photographs, and we include some of the more dramatic photographs of biomass burning. Glasser and Lulla (1997) point the way for future research in observing that the OES data "could be made to more closely represent actual biomass burning if one were to take into account information related to the picture content and to multiple exposures" (Glasser and Lulla, 1997:6). The photographs contain spatial and temporal information on biomass burning which has yet to be extracted and which we leave for subsequent studies. For example, some photographs show single fires, but some show tens or even hundreds of smoke plumes, each related to a fire. In broad oblique views of large areas, some smoke plumes may go unrecorded in the feature field of the database because catalogers generally enter the central location or feature of the photograph into the feature field of the OES database. A simple count of the number of photographs with biomass burning observed is misleading because crew members commonly shoot more than one view of the same scene. The same burning may also be photographed at multiple exposures, from different angles, and on subsequent orbital passes.

7.3 GLOBAL PATTERNS OF BIOMASS BURNING AND SMOKE PALLS

The spatial and temporal distribution of large-scale biomass burning has been documented in imagery from both robotic satellites and from human-occupied spacecraft, especially since the late 1970s (Helfert and Lulla, 1990; Lulla et al., 1994). Andreae's (1993) global survey showed that burning occurred in South Asia, Africa and Madagascar, Indonesia, portions of Australia, and the Americas; furthermore, burning in the northern hemisphere was most common in the spring and less so in the fall. In contrast, Andreae (1993) concluded that rain forest and savanna were burned in the southern hemisphere primarily in late winter and early spring (August and September). We examined the differences in spatial and temporal patterns since May 1992 (the end of Andreae's study).

7.3.1 Geographical Distribution of Burning and Smoke

Based on 1030 photographs of burning worldwide from May 1965 to May 1992, Andreae's (1993) map (Figure 7.1, top panel) showed many fires in the well-known zones of deforestation along the margins of the Amazon rain forest and in the Sahel of West Africa. But Andreae (1993) expressed surprise that large burning events were common in so many other parts of the world—specifically in Mexico and Cen-

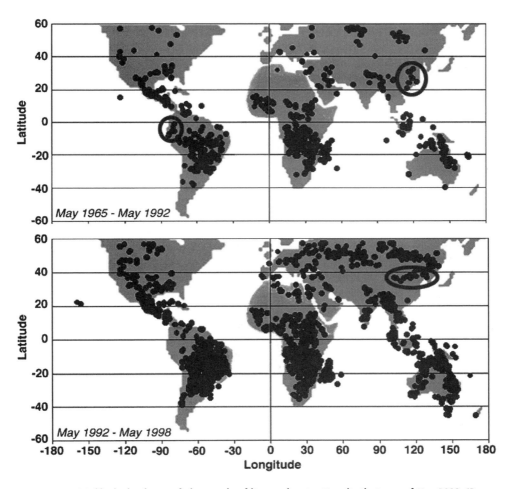

Figure 7.1 Worldwide distribution of photographs of biomass burning. Top: distribution as of May 1992. (Reconstructed from original data following Andreae, 1993; used with permission of the author and *Eos, Transactions of the American Geophysical Union.*) Bottom: geographic distribution of biomass burning photographs collected between 1992 and 1998. Approximate continent boundaries are provided for reference. The circled data are discussed in the text.

tral America, the grasslands of southern Brazil, Paraguay, and Argentina, as well as in the Andean countries of Bolivia, Peru, and Ecuador. The scale of burning south of the central African rain forest, representing "the largest concentration of fires in the world" (Andreae, 1993:129), had also been unsuspected.

High Northern Latitudes (North of 40°N). Between May 1992 and May 1998, the global distribution of burning observations showed significant differences (compare panels in Figure 7.1), mainly because of more frequent high-inclination Space Shuttle missions and the 51.6° inclination of the *Mir* spacecraft. These inclinations permitted observations from 40°N to a maximum of about 60°N in oblique views (especially from 57°-inclination Space Shuttle missions).

Since May 1992 we have many more observations of burning in the northern hemisphere north of 40°N in the boreal forest zone, particularly in Russia west of the Ural Mountains, in most of Siberia, and in North America (Figure 7.1).

Lower Latitudes (40°S to 40°N). The data acquired since mid-1992 show a general filling out of all continental surfaces that support burnable biomass, so that the map depicts burning of most major terrestrial biomes. Our additional data (Figure 7.1, bottom panel) show that the zone of fires in Eurasia north of 40°N is bordered to the south by an identifiable region without recorded fires, centered on latitude 40°N. This zone coincides with the arid Gobi and Takla Makan Deserts of northern and northwest China and extends south into the arid regions of Iran, Afghanistan, Pakistan, and northwestern India. Other deserts also show little incidence of biomass burning, including the Sahara Desert, the Arabian peninsula, and central and east Asian deserts in the northern hemisphere; in the southern hemisphere the Namib Desert in southwest Africa, Patagonia in southern South America, and the deserts of central Australia are rarely sites of burning.

The largest regions of burning in the southern hemisphere are the dense woodlands and savannas south of the Congo and Amazon rain forests. Our data increase the density of Andreae's (1993) fields and expand them southward. Fires have been mapped throughout the rain forests of the Congo basin and in most of the Amazon basin, except in the remote northwest and neighboring upper Orinoco basin. The remnant forests of the Eastern Ghats of India constitute a newly identified cluster of biomass burning.

Figure 7.1 (bottom panel) also shows that the scatter of data points expands in the western United States south of 40°N, and a significant cluster extends into northern Mexico. This expansion may result from the greater amount of data, showing burning even in semiarid regions. Other semiarid regions with new data points include the Sahel and East Africa, the Mediterranean basin, and the more densely vegetated and populated parts of Australia. The denser distribution of biomass-burning photographs shows a comparative lack of burning in the economic core regions of the world—northeastern United States, western Europe, and East Asia. Burning is recorded in a major new cluster in northern China, following the semiarid regions stretching across China for 2000 km (Figure 7.1, lower panel, circled). However, no new records of burning have appeared in eastern and southeastern China since Andreae's (1993) study (Figure 7.1, bottom panel), although Andreae (1993) identified a cluster in this region (Figure 7.1, top panel, right circle). The apparent lack of fires in eastern and southeastern China may reflect the 94% reduction in forest areas damaged by fire reported for the decade 1987 to 1997, a result of fire prevention efforts by the Chinese government (*International Forest Fire News*, 1997). Another area where burning was documented by Andreae (1993), but was not identified since, is within Peru (Figure 7.1, top panel, left circle).

7.3.2 Seasonal Distribution of Biomass Burning

Andreae (1993) showed global biomass burning plotted by date and latitude (Figure 7.2, top panel), which revealed distinct seasonality in biomass burning in many parts of the world. Andreae (1993) concluded that northern hemisphere burning was concentrated in the spring and to a lesser extent in the fall. In contrast, rain forest and

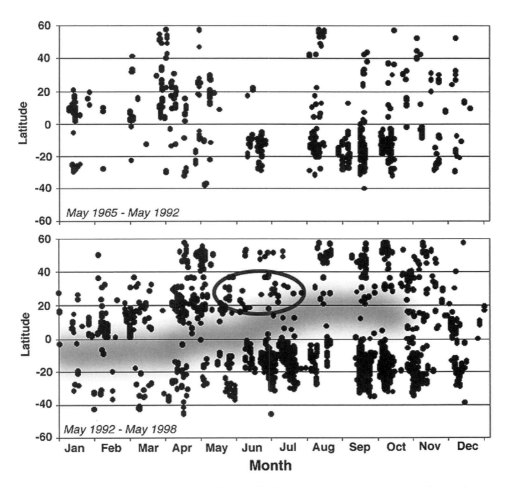

Figure 7.2 Global biomass-burning photographs plotted by date and latitude. Top: seasonality in biomass burning is evident in many parts of the world as identified by Andreae (1993) as of May 1992. (Reconstructed from original data following Andreae, 1993.) Bottom: data for biomass-burning photographs collected between 1992 and 1998. Shaded swath marked by fewer data points in the equatorial zone is discussed in the text. The swath moves across the equatorial zone in step with the seasonal shift of the ITC. The circled data points are also discussed in the text.

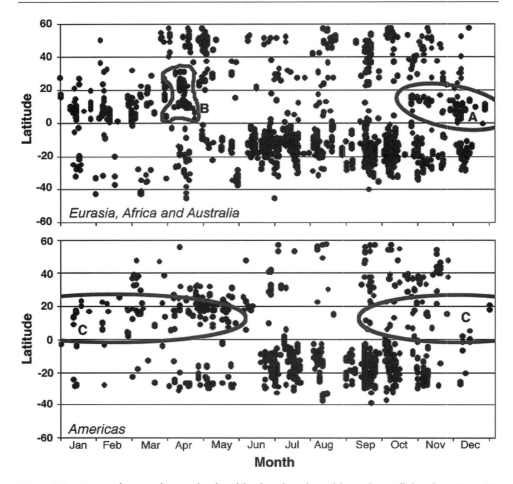

Figure 7.3 Biomass-burning photographs plotted by date, latitude, and hemisphere. All data from Figure 7.2 (1965–1998) were combined for this analysis, then separated by hemisphere to facilitate geographic interpretation. The regions marked A, B, and C are discussed in the text.

savanna were burned in the southern hemisphere primarily in late winter and early spring (August and September, Andreae, 1993). Andreae (1993) also discussed the seasonality of burning in smaller geographic regions (even in parcels as small as individual countries) when biomass-burning photographs happened to fall in clusters. We performed a similar date–latitude analysis on new data acquired since May 1992 (Figure 7.2, bottom panel), and on all combined data May 1965 to May 1998 (Figure 7.3).

High Northern Latitudes (North of 40°N). One of the most marked changes in data coverage since Andreae's (1993) study is the poleward increase in photographs of burning in the spring and fall (compare panels in Figure 7.2). Plotting the data points for Eurasia separately from those for the Americas shows that burning appears to begin in Eurasia in March (Figure 7.3, top panel), some months earlier than in North America, where the season seems to begin in June or July (Figure 7.3, bottom panel).

Lower Latitudes (40°S to 40°N). A new cluster of data points appears on our scattergram filling a gap between 10 and 40°N in the months May through July (Figure 7.2, bottom panel, circled). Although not as dense a cluster as others on the plot, these points represent fires in the latitudinal zone of Mediterranean climates of both the North Africa–Middle East region, and North America (Figure 7.3, top and bottom panels). At minimum, this new cluster provides baseline data for future investigations of burning seasonality at these latitudes.

Although we cannot interpret relative burning frequencies from the nonnormalized data, the distribution of points supports prior results about the seasonality of burning; the lower density of data points between the equator and 40°N in May through August probably is indicative of reduced burning in high summer in Eurasia (Figure 7.3, top panel). Similarly, burning appears to be reduced in Central and North America (as far north as 40°N) during summer (Figure 7.3, bottom panel).

A major feature that now appears on the date–latitude plot, and which is not discernible in Andreae's (1993) plot (Figure 7.2, top panel), is a distinct, burn-free swath in the equatorial zone. The swath moves across the equatorial zone in step with the seasonal shift of the Intertropical Convergence (ITC) zone in the atmosphere (shaded area, Figure 7.2, bottom panel). At a global scale, the ITC is related to the axis of equatorial rainfall maxima, with rainfall "following the sun" southward in southern summers and northward in northern summers. This ITC-related swath straddles about 20° of latitude in the months of February, May, and June. From July onward the zone of fewer fires is less marked but is still identifiable; the zone narrows and becomes poorly defined with the shift of the ITC southward between September and December.

There are reasons for suspecting that the pattern of reduced burning in regions with ITC rainfall is real, despite the fact that photographing equatorial land surfaces is hampered by ITC cloud masses. Crews have been trained and advised during flight to photograph equatorial regions whenever forecasts from satellites indicate that clouds are clearing. In addition, regardless of flight path inclination, the Space Shuttle and *Mir* always fly over the ITC, so there are numerous photographs of the high-rainfall equatorial zone in the database (Figure 7.1). It is well established that the intensity of burning in the equatorial zone is diminished during rainy seasons. The fact that the burn-free pattern follows the seasonal march of ITC rainfall (Figure 7.2, bottom panel) seems reasonable.

The obverse of establishing the lower incidence of biomass burning in the high-rainfall ITC zone is the greater definition of burning beyond this zone in the tropics (between about 25°N and S). We were able to define a burning season in Southeast Asia, particularly the beginning of the burning season in late October (Figure 7.3, region A). This cluster of points does not appear per se in Andreae's (1993) data (Figure 7.2, top panel). A related issue is how the equatorial fire seasons change during ENSO years. Although we do not present our ENSO versus non-ENSO analyses here, the plots of fires in ENSO years generally reinforce existing patterns. However, an analysis of Southeast Asian burning photographs from ENSO versus non-ENSO years reveals that the cluster of photographs in April between the equator and 25°N (Figure 7.3, region B) reflects a high density of smoke photographs during ENSO years. Burning late in the dry season is presumably a result of ENSO-related drought.

The few photographs of biomass burning in late December and early January reflect the rarity of Space Shuttle flights during that period. Glasser and Lulla (1997) have calculated for the Shuttle program through 1995 that photographs taken in

December represent only 1% of the total number of biomass-burning photographs, and in January only 3%. The greater continuity of the Shuttle–*Mir* data has begun to fill this gap (compare panels in Figure 7.2); observations from the International Space Station (ISS) will continue the trend.

This observation on flight scheduling and the clustering of data points provides a new perspective on burning in the low-latitude Americas (Figure 7.3, bottom panel). On the basis of the data acquired up to May 1992, burning in the equatorial zone was specifically noted by Andreae (1993) for Central America (0 to 20°N) in January and May (Figure 7.2, top panel). Based on the greater amount of data available now and the low counts of biomass-burning photographs for December and January due to flight scheduling, we conclude that a discrete fire season probably begins much earlier in October (Figure 7.3, region C). We see the burning season as more continuous from October through June, with a peak in April and May, before the wet season suppresses burning from July through September.

We note an expansion of data points in the southern hemisphere. Andreae's (1993) cluster centered at 10°S in June (Figure 7.2, top panel) is here shown to be part of a longer burning season in the southern hemisphere, with burning starting in March and reaching a long crescendo between June and October, at latitudes between the equator and 35°S (Figure 7.2, bottom panel). South America, Africa, and Australia all contribute to this cluster (Figure 7.3). Burning in low to midlatitudes of South America starts as early as March. This is a result of the beginning of the dry season, as indicated by the large, dense cluster of observations centered on 20°S from April through early November (Figure 7.3, bottom panel). Gaps in this cluster (early June and late August) probably do not indicate real hiatuses in the southern burning season, as the dry season in south-central Brazil is continuous from May through October.

Andreae (1993) and subsequent researchers have noted the complex pattern of the burning south of the Equator in Africa. Although related to the dry winter when the ITC is farthest from south-central Africa, the spatial progression of burning documented by Andreae (1993) and Cahoon et al. (1992) is confirmed even by these nonnormalized data. When data points for Africa are mapped (not shown here), the burning season can be seen to begin around May in Angola and the Democratic Republic of Congo (former Zaïre) and to shift eastward in the course of the dry season until November and December, when southeast Africa and Madagascar experience peak burning rates.

Andreae (1993) had concluded, based on fewer data points, that a relationship between the onset of the dry season and burning in the low latitudes is simplistic. The shift of burning in south-central and southern Africa from west to east is confirmed by our data; that pattern is superimposed on the seasonal shift of the larger ITC feature. On the basis of these data, the several examples illustrated above, and our analysis of datelined clusters of data points, we concur with Cahoon et al. (1992) that the beginning of the dry season is indeed a major control of the seasonality of biomass burning in the low latitudes.

7.4 EXAMPLES OF PHOTOGRAPHS OF REGIONAL BIOMASS BURNING

Shuttle–*Mir* and other photographs of burning are presented for four significant regions: the Russian Far East–China–Mongolia region, Southeast Asia, Mexico–Central America, and south-central Africa.

7.4.1 Russian Far East, China, and Mongolia

Very large fires were recorded in China and the Russian Far East in 1987. Immense areas are covered by forests in East Asia; maps of the coniferous forests of Eurasia show that forests of the Russian Far East alone cover more than 3.3 million square kilometers, more than twice the area of the Amazon rain forest (Britannica Online, 1998). These facts have attracted attention to the boreal forests in the last decade. It is estimated that the former Soviet Union lost 12% of its forests between 1850 and 1980, and that China lost 39% in the same period (International Institute for Environment and Development and World Resources Institute, 1987). We show here what may be the first published views of very large smoke palls from the Russian Far East, China, and Mongolia.

At the start of NASA 2 in March 1996, the central regions of Asia, including southern Russia, northern Mongolia, and to a lesser extent, northeastern China, were experiencing wildfires (*International Forest Fire News*, 1996). The weather patterns of central Asia engender two distinct fire seasons. In spring, dry forests and steppe grasses are subjected to strong, dry, northwesterly winds when the main fire season occurs. After wetting by rains in the summer months, a weaker fire season occurs in October and November when the vegetation has again dried out (shown in Figure 7.3). The spring burning season of 1996 followed the unusually dry winter of 1995–1996, when the taiga forest zones and steppe vegetation were tinder dry. The most favorable fire situation occurred in the eastern part of the Russian Federation, including the Krasnoyarsk, Baikal, and Yakutiya regions. Of the 1.8 million hectares that burned, there were more than twice as many fires in the large class (fires covering more than 200 ha) than in previous years, and these large fires burned 1.6 million hectares, 89% of the area burned. In combination, burning occurred on a scale unprecedented in the recollections of fire officials (*International Forest Fire News*, 1996).

Similarly, Mongolia began to experience wildfires in March and April 1996 (Figure 7.4; see the color insert) that were significantly greater than would normally be expected for that season (*International Forest Fire News*, 1996). Approximately 15.6 million hectares of the vast Siberian taiga forest grows in northern Mongolia (much of which is shown in the broad views of Figures 7.4B and 7.5A; see the color insert). Of a total of 386 fires that occurred during the 1996 spring fires, 36 were uncontrolled. Eventually, 2.3 million hectares of forested land as well as 7.8 million hectares of steppe grassland south of the taiga forest were burned. One fire, which extended along a 129-km front, migrated from Mongolian into Russian territory. According to local authorities, the primary cause of forest and grassland fires in Mongolia is related to careless disposal of cigarettes. Both Russia and Mongolia were hampered in their efforts to suppress wildfires because of economic constraints. The increase in number of fires was correlated to the reduction in funding available to fight them: In 1996, aircraft flying time devoted to the surveillance of fires was reduced from 150,000 hours to 40,000 hours (*International Forest Fire News*, 1997).

Many areas in the Russian Far East and Mongolia had large fires again in 1997 and 1998. In late April and early May 1997, crews on *Mir* recorded hundreds of fires and a regional smoke pall across Mongolia, eastern Russia, northern China and Korea, and out over the Pacific as far as Japan (Figure 7.5). The pattern of fires suggested that they were the result of human action. The Office for the Coordination of Humanitarian Affairs of the United Nations (OCHA) (1998) reported that people in these regions were foraging more widely into forests as the economic situation declined in the Russian Federation and Mongolia. The fires and smoke were recorded for about two weeks by the astronauts and cosmonauts but were poorly documented in the news media.

Fires in the Russian Far East were environmental problems in 1998 as well. OCHA (1998) reported 1028 fires in 1998 in the relatively small Khabarovsk Krai (administrative unit) alone, of which 18 were very large, each burning between 20,000 and 30,000 hectares of forest. This situation led authorities to declare a state of emergency in July 1998.

Meteorological satellite data are insufficiently detailed to provide early warning of Siberian fires in the early stage when fires are small (OCHA, 1998); combined with the fact that surveillance by aircraft has been curtailed in Russia's heavily forested Siberian and Far Eastern regions, it is apparent that there is a potential practical application for digital imagery taken from cameras in low Earth orbit. A similar potential may exist for surveillance of new fires from low Earth orbit in other parts of the world.

7.4.2 Southeast Asia

Continent-scale smoke palls from wildfires occurred in Southeast Asia starting in the spring of 1997. By September 27, 1997, the Southeast Asian smoke pall associated with fire in Indonesia stretched continuously for more than 4000 km from east to west (Total Ozone Mapping Satellite data, Figure 7.6; see the color insert). Air quality monitors in Singapore and Malaysia documented record levels of smoke (Simons, 1998). The fires and associated smoke were recorded from September through November 1997, by crew members on the *Mir* station and Space Shuttle mission STS-86. Astronauts reported that they were unable to see the land or ocean surfaces in the Indonesian archipelago or in neighboring Malaysia. An image mosaic (Figure 7.6) shows smoke emitted from fires over Sumatra. Smoke from many wildfires can be seen adding to the extensive smoke pall that covered Indonesia and the surrounding region at the time.

The tropical forests of Indonesia rank third in area after Brazil and Democratic Republic of Congo, and cover an estimated 0.9 to 1.2 million km^2 (*International Forest Fire News*, 1998). Preliminary estimates of the extent of forest burned by October 1997 indicated that approximately 1650 km^2 of forest burned, and up to 20,000 km^2 was burned when all vegetation associations are considered (*International Forest Fire News*, 1998). An analysis completed in southern Sumatra in December 1997 suggested that more than 27,980 km^2 of vegetation, including 7010 km^2 of forest, burned on the island of Sumatra alone (*International Forest Fire News*, 1998).

In Indonesia, burning is almost entirely an anthropogenic process. Whether employed to clear agricultural land of refuse or to clear forested lands for other uses, it has been part of the way of life in the region for millennia (*International Forest Fire News*, 1998). For many rural people, fire is the only tool for preparing land for subsistence farming. Corporations use fires to clear forested land for plantations and other uses (Simons, 1998). Farmers and foresters took advantage of the drought conditions to clear land that previously had been too wet to be burned successfully. Some fires escaped into the drier forests and tree plantations as wildfires.

Major wildfires occurred in Indonesia in 1982–1983, 1987, 1991, and 1994 in connection with the El Niño weather cycle (*International Forest Fire News*, 1998). The El Niño event of 1997–1998 was more intense and persistent than usual. It was comparable to the 1982–1983 El Niño, which also contributed to dramatic interregional smoke palls and resulted in the greatest area of forest burned, 24,000 to 36,000 km² in east Kalimantan alone. The El Niño weather cycle characteristically brings drought to the Australasian region early in the year, but in 1982 and 1997 it persisted into the following year, delaying the wet season of the monsoon cycle, which normally dampens fires after September (*International Forest Fire News*, 1998). In Indonesia, the early onset and persistence of the drought led to extreme water shortages, crop failures, and famine. Much of the smoke was generated when water tables dropped and allowed deep smoldering fires to burn in the organic soils and peat lands of coastal regions in Sumatra and Kalimantan.

Special meteorological factors were important in distinguishing the 1997 fire season from previous years and appear to have produced more smoke. A capping inversion related to the unusual high-pressure system prevented upward diffusion of smoke, thereby contributing to record concentrations of smoke (*International Forest Fire News*, 1998). Because of the international tensions that arose as large smoke clouds moved across national boundaries, a call was made for political action to modify human behavior, the major cause of burning in Indonesia (*International Forest Fire News*, 1998).

7.4.3 Mexico–Central America, April–May 1998

In spring 1998, smoke from Mexico and Central America spread north to the southern United States. At its peak, air pollution monitors in Brownsville, Texas, recorded mass loading levels of 500 µg m^{-3} (Texas Natural Resource Conservation Commission, unpublished data; see also Glasser and Lulla, 1998a). Smoke from these fires was photographed by the crews of NASA 7, STS-90, and STS-91. These photographs captured the development of the fires from their beginning in April through the peak of the episode in May 1998 (Figure 7.7; see the color insert). The densest smoke in the photographs came from clusters of large multiple fires located on the higher forested slopes on the Pacific side of the Sierra Madre del Sur.

Factors responsible for the wildfires were similar to those in Indonesia and other tropical regions undergoing rapid deforestation. Fires in Mexico and Central America were started by both large and small landowners to clear vegetation and burn wastes on agricultural lands, and to clear forests. In the dry conditions of the 1997–1998 El Niño, fires escaped into forested areas which in wetter years would not have caught fire.

The effect of the ENSO-related droughts was also apparent from other satellites (Figures 7.6 to 7.8). In Figure 7.7, the TOMS satellite aerosol index provides a measure of the density of smoke penetrating into the southern United States and over the Caribbean and Pacific oceans, more than 2500 km from the source fires. Figure 7.8 is a compilation of nighttime data observations of Central American burning in visible–near infrared wavelengths obtained by the Defense Meteorological Satellite Program (DMSP) and compares a normal burning year (1993) with the 1998 season (National Oceanic and Atmospheric Administration, 1998). The comparison shows a dramatic increase in numbers of fires across the region in 1998. The international impact of large biomass burning events is well illustrated in both types of orbital views (Figures 7.7 and 7.8).

7.4.4 South-Central Africa

South-central Africa supplies significant quantities of aerosol and trace gases to the atmosphere (Hao et al., 1990; Andreae, 1991; Lacaux et al., 1993). Biomass burning in the African savannas (which stretch from the Congo River basin to southeast South Africa) generates a biogenic haze of aerosols with high concentrations of such trace gases as CO, NO_x, and CH_4Br (R. Swap, University of Virginia, Char-

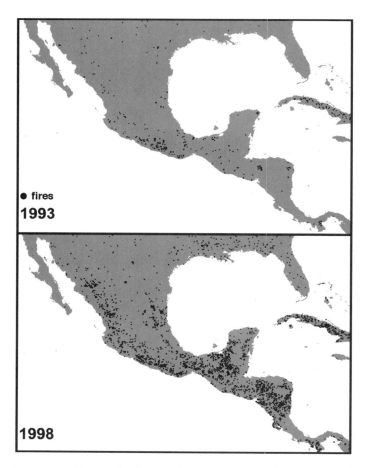

Figure 7.8 Fires (dots) in Mexico and Central America in the burning seasons of 1993 and 1998. The data were obtained from the DMSP Operational Linescan System (National Oceanic and Atmospheric Administration, 1998). To enhance the contrast, we expanded the area of each dot by one pixel for both maps.

lottesville, personal communication, 1998). Figure 7.9 depicts an increasing intensity of burning northward, as savanna biomass volume rises in response to higher rainfall toward the equator. Photos from the Space Shuttle and Shuttle–*Mir* programs have regularly illustrated the number of individual fires and the extent of resulting large smoke palls, which have covered 150,000 to 250,000 km^2.

It is estimated that fires in tropical savannas burn three times more biomass each year than do fires in tropical forests (Andreae, 1991). Figure 7.9 shows a sharp boundary along the ecotone between savanna and rain forest, indicating that biomass burning of the Congo rain forest is low compared with that in neighboring savannas with high flammable biomass volumes. However, Figure 7.9 is based on two years of data (1986 and 1987) derived from nighttime low-light imagery from block 5 satellites of the Defense Meteorological Satellite Program (DMSP; Cahoon et al., 1992). The longer record of the Space Shuttle and Shuttle–*Mir* photograph database provides a different perspective: namely, that fires have been widely distributed in the Congo rain forest (and much of the Amazon and other rain forests) in the course of one to two decades (Figure 7.1).

In Figure 7.10 we show photographs of biomass burning in Angola, one of the most fire-prone countries in sub-Saharan Africa. The boundary between low and high fire frequency accompanies the ecotone between arid savanna margins in southern Namibia and wetter savannas in southern Angola (Figure 7.10A). Since southern Africa supplies a significant amount of aerosol to the atmosphere in the southern hemisphere (Hao et al., 1990; Andreae, 1991; Lacaux et al., 1993), it is not surprising that

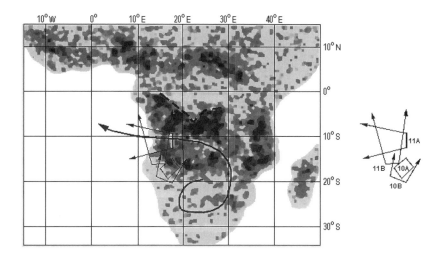

Figure 7.9 Distribution of seasonal forest fires in Africa and Madagascar (shaded areas), as sensed by DMSP satellite sensors in 1986 and 1987. The boundary between the rainforest biome and the savanna woodlands, which are more prone to burning, is shown by a dashed black line. The density of fire distribution wanes southward with the decline in rainfall in the savanna and concomitant decline in biomass. Anticyclonic circulation of the atmosphere over southern Africa (gray arrow) transports smoke typically via Angola into the tropical South Atlantic Ocean. Areas and view directions of the four photographs (Figures 7.10 and 7.11) are indicated with thin black lines. (Modified from Cahoon et al., 1992; Garstang et al., 1996.)

the largest CO_2 and low-level ozone anomaly (2 to 5 km above sea level) in the southern hemisphere lies immediately downwind in the tropical South Atlantic Ocean, or that the core of the anomaly partly overlies Angola (Fishman, 1992) (Figure 7.10B). Swap et al. (1996) have shown that 135 Mt/yr of aerosol is delivered into the atmosphere over southern Africa, some of which feeds the low-level ozone anomaly.

The significance of this region as a supplier of particulates to the atmosphere can also be judged from the albedo effect of smoke palls on models of regional climates. In recent models describing the cooling effect of aerosols on the atmospheric heat budget, the three largest aerosol-dominated anomalies are all in the northern hemisphere and are derived mainly from industrial and urban hazes (northeastern United States, western Europe, and East Asia) rather than biomass burning hazes. However, the largest center of aerosol-related cooling in the southern hemisphere lies over south-central Africa, especially Angola (Kiehl and Briegleb, 1993; Charlson and Wigley, 1994).

Researchers on air-mass evolution in South Africa (Andreae, 1991; Andreae et al., 1996; Garstang et al., 1996) have recently described aerosol sources, air-mass residence time, trajectories, and weather systems by which hazes are ultimately trans-

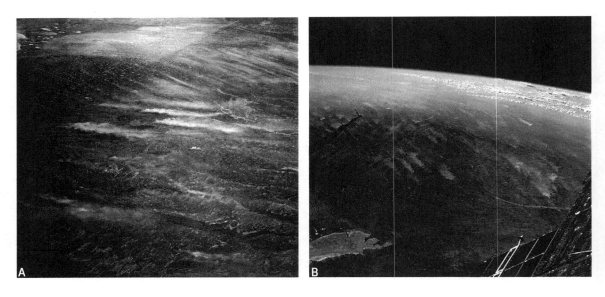

Figure 7.10 Angola–Namibia borderlands, August 5, 1997 (locations mapped in Figure 7.9). (*A*) In this southwest-looking view (NASA photograph NASA5-707-14), the linear east-west border between Angola and Namibia appears prominently across the top of the view, with light tones indicating significantly thinner vegetation on the Namibian side (top left). By contrast, numerous fires appear in the savannas of southern Angola (center), where biomass volume is greater than in Namibia (due to decreasing rainfall, higher human population densities, and heavier grazing by herd animals on the Namibian side). As indicated by the smoke plumes from individual fires, two wind directions are apparent: in the top half (to the west), winds are southwesterly but in the lower half, winds are from the south-southeast. (*B*) This north-looking view (NASA photograph NASA5-706-196) shows the dry lake bed of Etosha Pan in northern Namibia (lower left) and many fires in the middle ground. A large, diffuse smoke pall in central Angola with a marked southern boundary stretches across the top of the view. Clouds appear on the horizon, top right, above the smoke layer. This more oblique view complements view (*A*) by showing the same fires in the wider context of regional burning.

ported from the subcontinent (see Chapter 6). Some weather systems transport hazes offshore fairly directly. Other weather systems with recirculating winds maintain the hazes onshore, incorporate industrial pollutants from the South African industrial core, and supply the capping mechanism that holds aerosols near the ground. This combination increases the measured aerosol concentrations and also the visibility of aerosol masses from low Earth orbit (Garstang et al., 1996). Air masses can recirculate onshore for weeks. Under clear skies, chemical evolution of these hazes can proceed before the air mass is transported off the subcontinent in this region via the Angola "exit" into the tropical Atlantic Ocean (Figures 7.9 and 7.11). A NASA-sponsored series of experiments (SAFARI 92 and SAFARI 2000; Andreae et al., 1996) continue to use ground-based, airborne, and satellite observations to investigate the source regions of aerosols and trace gases on both sides of the tropical South Atlantic Ocean.

7.5 CONCLUSION

The spatial and temporal distribution of large-scale biomass burning has been documented in images from robotic satellites and human-occupied spacecraft only in the last 20 years (Helfert and Lulla, 1990; Lulla and Helfert, 1989). The extension of the database of astronaut photographs of biomass burning since 1992 allowed us to revisit Andreae's (1993) global analysis of biomass burning. We show differences in spatial distribution of observations acquired after Andreae's (1993) study.

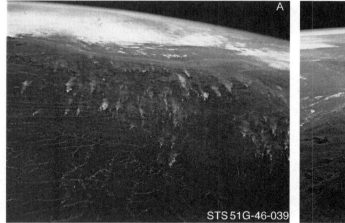 

Figure 7.11 Transport of smoke to the Atlantic Ocean, June, 19, 1985 (locations mapped in Figure 8.9). (*A*) This oblique photograph of Angola in the dry season looks west toward the Atlantic Ocean and shows individual fires in the savannas (NASA photograph STS51G-46-39). Smoke plumes indicate wind movement to the west (bottom to top). (*B*) This is one of the best photographs illustrating the transport of smoke and pollutants from urban South Africa into the tropical South Atlantic (NASA photograph STS51G-46-77). Individual plumes give the sense of airflow toward the Atlantic Ocean.

Our map of global biomass burning constructed from the photographs of the planet since May 1992 shows biomass burning in all parts of the world except Antarctica and the arid desert cores. The most striking result is the increase in the number of photographs recording burning in the entire northern hemisphere between 40 and 60°N, which includes the high-latitude forests of Russia and North America. There are sufficient data to suggest that seasonal biomass burning begins in Eurasia in March, some months earlier than in North America, where the season appears to begin in June or July. In addition to the records of fires in Asia in the far northern latitudes, our data define significant biomass burning in northern China. The trends we have suggest that additional attention be given to these regions in future studies.

Our date–latitude analysis of all photographic data available through May 1998 provides a baseline for more detailed investigations of the relationship between peak burning and dry seasons, especially when treated geographically by region or subcontinent. We made new estimates of season of burning for a number of areas, including North America, northeast Asia, the dense woodlands and savannas south of the Congo and Amazon rain forests, and Central America. We also identified a swath of suppressed biomass burning that follows the seasonal shift of the Intertropical Convergence Zone (ITC). The shift of burning from west to east in south-central and southern Africa during the dry season was confirmed by the new data and appears to be a pattern superimposed on the seasonal movement of the ITC. Finally, plots of fires in ENSO years generally reinforced existing patterns of burning distribution and seasonality, except for an increase in the number of late-season observations in Southeast Asia.

Astronaut photographs are valuable in assessing the occurrence and extent of smoke from fires around the world. The regional transport of smoke across international boundaries is an issue of increasing concern, and the potential applications of these graphic images of international smoke palls include the assessment of smoke boundaries, optical thickness, and detailed changes over time. For example, astronauts and cosmonauts documented several international smoke events resulting from wildfires in 1996–1998 (far northeastern Asia, Southeast Asia, and Central America) and obtained the first photographs of smoke and pollutant movement from south-central Africa to the tropical South Atlantic Ocean. Also, the increased use of real-time digital imaging by astronauts and cosmonauts on the Space Shuttle and International Space Station and a continuous presence in space will add astronaut photography to the methods of wildfire surveillance, especially in remote regions.

Finally, we note that the photographs in the Office of Earth Sciences (1999) database contain unused spatial and temporal information on global biomass burning. The continent-wide perspectives provided by astronauts in broad oblique views can be exploited in research on biomass burning in other parts of the world. Conversely, detailed photographs from low Earth orbit also provide sufficient resolution for local patterns of burning to become apparent, as in the case of fires in the margins of forests in southern Siberia and northern Mongolia (Figure 7.5). Furthermore, photographs from low Earth orbit illustrate the meteorological conditions prevailing during biomass burning events and provide an excellent tool for integrating other meteorological and air-chemistry data (see Chapter 6).

Nearly all global climatic models of greenhouse warming indicate that warming should appear first and be most pronounced in the northern latitudes (Schneider,

1996). If warmer and drier conditions occur in the extensive boreal forests, these regions will probably become more susceptible to fires. Thus, data from high-latitude Shuttle–*Mir* missions provide baseline data for comparison with future photographs of these regions from the International Space Station (Glasser and Lulla, 1998b).

ACKNOWLEDGMENTS

We thank members of U.S. and Russian crews for photographing numerous biomass burning events, often under conditions that were not optimal. We also thank C. A. Evans and J. A. Robinson for discussions and editorial comments in preparation of the manuscript. J. A. Robinson performed the database analysis and plotted the data, and L. Prejean completed the figures with expertise and speed. C. A. Evans constructed the mosaic of photographs over Indonesia. We thank Jay Herman for information concerning interpretation of the TOMS satellite data.

REFERENCES

Andrasko, I., Ahuja, D. R., and Tirpak, D. A. 1990. Policy options for managing biomass burning to mitigate global climate change. *Abstract, Chapman Conference on Global Biomass Burning: Atmospheric Climate and Biospheric Implications,* Williamsburg, VA, March 19–23.

Andreae, M. O. 1990. Biomass burning in the tropics: impact on environmental quality and global climate. *Proceedings, Chapman Conference on Global Biomass Burning: Atmospheric, Climatic, and Biospheric Implications,* Williamsburg, VA, March 19–23, Vol. 1., Chap. 1.

Andreae, M. O. 1991. Biomass burning: its history, use and distribution and its impact on environmental quality and global climate. In Levine, J. S., ed. *Global Biomass Burning: Atmospheric, Climatic, and Biospheric Implications.* Cambridge, MA: MIT Press, pp. 3–21.

Andreae, M. O. 1993. Global distribution of fires seen from space. *Eos, Transactions of the American Geophysical Union,* 74:129–130.

Andreae, M. O., Fishman, J., and Lindesay, J. 1996. The Southern Tropical Atlantic Region Experiment (STARE): Transport and Atmospheric Chemistry near the Equator-Atlantic (TRACE A) and Southern African Fire-Atmosphere Research Initiative (SAFARI): an introduction. *Journal of Geophysical Research,* 101(D19):23519-23520.

Booth, W. 1989. Monitoring the fate of the forests from space. *Science,* 243:1429.

Botkin, D. B., Caswell, M., Estes, J., and Orio, A., eds. 1989. *Changing the Global Environment: Perspectives on Human Involvement.* New York: Academic Press, 459 pp.

Britannica Online. 1998. Figure 53. Worldwide distribution of boreal forests. *http://www.eb.com:180/cgi-bin/g?DocF=cap/abiosph073m4.html* (viewed December 21, 1998)

Cachier, H. 1992. Biomass burning sources. In W. A. Nierenberg, ed., *Encyclopedia of Earth System Sciences,* Vol. 1. New York: Academic Press. pp. 377–385.

Cahoon, D. R., Stocks, B. J., Levine, J. S., Cofer, W. R., and O'Neill, K. P. 1992. Seasonal distribution of African savanna fires. *Nature,* 359:812–816.

Charlson, R. J., and Wigley, T. M. L. 1994. Sulfate aerosol and climatic change. *Scientific American*, 270(2):48–57.

Fishman, J. 1992. Probing planetary pollution from space. *Environmental Science and Technology*, 25:613–621.

Garstang, M., Tyson, P. D., Swap, R., Edwards, M., Kållberg P., and Lindesay, J. A. 1996. Horizontal and vertical transport of air over southern Africa. *Journal of Geophysical Research*, 101(D19):23721-23736.

Glasser, M. E., and Lulla, K. P. 1997. Assessment of biomass burning using the NASA astronaut photography database, SSEOP. *Proceedings, American Society for Photogrammetry and Remote Sensing, 1997 ACSM-ASPRS Annual Convention and Exposition*, Seattle, WA, Vol. 3: *Remote Sensing and Photogrammetry*, pp. 333–342.

Glasser, M., and Lulla, K. P. 1998a. Astronaut's view of Mexico and Central America burning (April–May 1998). In *Earth Observations and Imaging Newsletter*. Houston, TX: Office of Earth Sciences, NASA Johnson Space Center. Available at: *http://eol.jsc.nasa.gov/newsletter/Mexico_Burning* (viewed March 2, 1999)

Glasser, M. E., and Lulla, K. P. 1998b. Evidence of change in the boreal forests using NASA astronaut photography: the case for International Space Station Earth observations. *Proceedings, 27th International Symposium on Remote Sensing of the Environment*, Tromsø, Norway, June 8–12, pp. 282–287.

Hao, W. M., Liu, M.-H., and Crutzen, P. J. 1990. Estimates of annual and regional releases of CO_2 and other trace gases to the atmosphere from fires in the tropics, based on the FAO statistics for the period 1975–1980. In Goldammer, J. G. ed., *Fire in the Tropical Biota: Ecosystem Processes and Global Challenges*. New York: Springer-Verlag, pp. 440–462.

Helfert, M., and Lulla, K. P. 1990. Mapping continental-scale biomass burning and smoke palls over the Amazon basin as observed from the Space Shuttle. *Photogrammetric Engineering and Remote Sensing*, 56:1367–1373.

International Forest Fire News. 1996 (September). No. 15. Geneva: FAO/ECE Agriculture and Timber Division, United Nations.

International Forest Fire News. 1997. (January). No. 17. Geneva: FAO/ECE Agriculture and Timber Division, United Nations.

International Forest Fire News. 1998. (January). No. 18. Geneva: FAO/ECE Agriculture and Timber Division, United Nations.

International Institute for Environment and Development and World Resources Institute. 1987. *World Resources, 1987*. New York: Basic Books, 369 pp.

Kiehl, J. T., and Briegleb, B. P. 1993. The relative roles of sulfate aerosols and greenhouse gases in climate forcing. *Science*, 260:311–314.

Lacaux, J.-P., Cachier, H., and Delmas, R. 1993. Biomass burning in Africa: an overview of its impact on atmospheric chemistry. In Crutzen, P. J., and Goldammer, J. G., eds., *Fire in the Environment: The Ecological, Atmospheric and Climatic Importance of Vegetation Fires*. New York: Wiley, pp. 159–192.

Levine, J. S. 1990a. Global biomass burning: atmospheric and biospheric implications. *Eos, Transactions of the American Geophysical Union*, 71:1075–1077.

Levine, J. S. 1990b. Atmospheric trace gases: burning trees and bridges. *Nature*, 346:511–512.

Levine, J. S. 1991. *Global Biomass Burning: Atmospheric, Climatic, and Biospheric Implications*. Cambridge, MA: MIT Press, pp. xxv–xxx.

Lulla, K., and Helfert, M. R. 1989. Space Shuttle Earth observations. *Geocarto International,* 4(2):67–80.

Lulla, K., Helfert, M. R., and Holland D. 1994. The NASA Space Shuttle Earth observations database for global change science. In Vaughan, R. A., and Cracknell, A. P., eds., *Remote Sensing and Global Climate Change.* NATO ASI Series, Vol. I, No. 24. Berlin: Springer-Verlag, pp. 355–365.

Lulla, K., Evans, C., Amsbury, D., Wilkinson, J., Willis, K., Caruana, J., O'Neill, C., Runco, S., McLaughlin, D., Gaunce, M., McKay, M. F., and Trenchard, M. 1996. The NASA Space Shuttle Earth observations photography database: an underutilized resource for global environmental geosciences. *Environmental Geosciences,* 3:40–44.

National Aeronautics and Space Administration, Goddard Space Flight Center, Atmospheric Chemistry and Dynamics Branch. 1998a. Earth probe TOMS: smoke over Indonesia for Sep 26, 1997.
http://jwocky.gsfc.nasa.gov/aerosols/indonesia/indo269.gif (viewed March 2, 1999)

National Aeronautics and Space Administration, Goddard Space Flight Center, Global Fire Monitoring. 1998b. TOMS aerosol daily data from Mexico, April 5–May 26, 1998.
http://modarch.gsfc.nasa.gov/fire_atlas/MEXICO/TOMS/mexico_toms_mov.html (viewed March 2, 1999)

National Oceanic and Atmospheric Administration, Defense Meteorological Satellite Program. 1998. Mexico and Central America fires: 1993 and 1998.
http://www.ngdc.noaa.gov/dmsp/fires/9398MexComp.html (viewed March 2, 1999)

Office for the Coordination of Humanitarian Affairs of the United Nations. 1998. Forest fires on the Island of Sakhalin and the Khabarovsk Krai: UNDAC (United Nations Disaster Assessment and Coordination) mission report.
http://www.reliefweb.int (viewed November 22, 1998)

Office of Earth Sciences, NASA Johnson Space Center. 1999. Office of Earth Sciences database of photographic information and images.
http://eol.jsc.nasa.gov/sseop (viewed March 2, 1999)

Pyne, S. J. 1995. *World Fire: The Culture of Fire on Earth:* New York, Henry Holt, 379 pp.

Schneider, S. H. 1996. Boreal forests. In *Encyclopedia of Climate and Weather.* New York: Oxford University Press, p. 360.

Simons, L. 1998. Plague of fire. *National Geographic,* 194(2):100–119.

Swap, R., Garstang, M., Macko, S. A., Tyson, P. D., Maenhaut, W., Artaxo, P., Kållberg, P., and Talbot, R. 1996. The long-range transport of southern African aerosols to the tropical South Atlantic. *Journal of Geophysical Research,* 101(D19):23777–23791.

Chapter

8

Windows of Opportunity: Photo Survey of the *Mir* Earth Observation Windows

Premkumar B. Saganti
Image Science and Analysis Group
NASA Johnson Space Center
Houston, Texas USA

Kamlesh P. Lulla
Office of Earth Sciences
NASA Johnson Space Center
Houston, Texas USA

ABSTRACT

Understanding the impact of the harsh space environment on windows used for Earth observations from *Mir* is important for planning scientific observations through windows on the International Space Station. Shuttle–*Mir* missions provided unique opportunities to document the current condition of the optical windows of *Mir* that have been exposed to the space environment for over a decade. We present results of the photo survey of windows in *Mir*'s Base Block, Kvant-2, and Priroda modules. These windows were used for most Earth observation activities by the astronauts and cosmonauts during Shuttle–*Mir* missions. The survey reveals some probable micrometeoroid and orbital debris impacts, translucency of window surfaces, bright spots that could have been the contamination deposits on the external panes, and other unique patterns on the window surfaces. It is conceivable that these impacts affect the quality of Earth observation imagery obtained through these windows. A qualitative description of selected window panes are provided with photo illustrations and image enhancements.

8.1 INTRODUCTION

One of the objectives of the Shuttle–*Mir* Earth sciences program was to understand the operational environment of long-duration missions, in preparation for more rigorous Earth sciences programs on the International Space Station (ISS). We used images of *Mir* windows of different ages to document the physical characteristics and degradation of the windows used for surveying Earth. The specific objectives of this photo survey of the *Mir* windows were (1) to gather data on the space environmental effects on *Mir* windows over time, and (2) to document the current condition of the windows used for Earth observation.

The oldest module of the *Mir* station (Figure 8.1, see the color insert) had been on orbit for 12 years (since 1986); the last module addition to the *Mir* was in 1996. The entire spacecraft contains more than 30 windows. Image data on eight different windows located in three separate modules (Table 8.1, modules shown in Figure 8.2) were collected by astronauts and cosmonauts over a period of 18 months during three successive Shuttle–*Mir* missions: NASA 5 (May 1997–September 1997), NASA 6 (September 1997–January 1998), and NASA 7 (January 1998–June 1998). Here we present results from analysis of images of five of the windows surveyed.

8.2 METHODS

8.2.1 On-Orbit Image Acquisition

Each window in our survey was imaged at several viewing angles, including at least one head-on view and several different oblique angles. Image data for both inner pane and outer pane of the specified windows were obtained under different lighting conditions. Interior illumination from the hand-held lights was used during dark portions of the *Mir* orbit to capture the blemishes on the inner window pane. A dark interior with bright exterior illumination (from either the ambient Earth-shine light or direct incident sunlight during terminator crossings) was used to capture the blemishes on the outer window pane.

Overall views of specific windows were acquired using a 35-mm Nikon-F3 film camera, with a wide field-of-view lens (35-mm setting on the 35/70-mm lens). Crew members also acquired close-up views of specific areas of anomalies on the window surfaces (scratches, smudges, impact cavities, and discoloration) using a zoom lens (55- to 70-mm setting on the 35/70-mm lens). Overall and close-up views of the window surfaces, window housing, and window surroundings were also obtained using a Cannon-L2 8-mm video camcorder to provide context and to facilitate the correlation of the detailed image data with other photography.

TABLE 8.1 Compilation of *Mir* Windows Surveyed in 1997–1998 Using Video and Film Media

Module	Year in Orbit	Windows Discussed in This Chapter (Total Number of Windows Surveyed)
Base Block	1986	Windows 1, 2, and 6 (6)
Kvant-2	1989	Window V (1)
Priroda	1996	Window 1 (1)

Methods 123

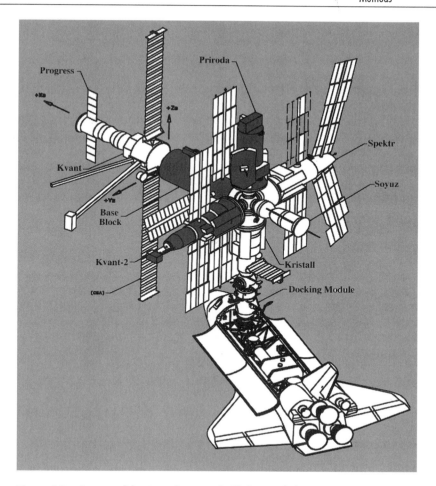

Figure 8.2 Diagram of the *Mir* configuration, highlighting in dark gray indicates modules containing windows in the survey: Base Block, Kvant-2, and Priroda.

8.2.2 Image Data Analysis

Photographic image data were translated into high-resolution digital scans for image analysis and image enhancement applications. All the 35-mm film image data were digitized from film to 4000 × 6000 pixels using a Kodak 4050 Professional Scanner. Using the Access software provided by Kodak, these high-resolution images were displayed on high-resolution screens and were visually analyzed. On these high-resolution images, image enhancement filters, including median filters and some customized filters, were used to generate three-dimensional intensity contour maps of micrometeoroid and orbital debris (MMOD) impacts. Window survey image data in video format was digitized to 480 × 640 pixels and was used for image analysis applications and to develop window orientation illustrations.

8.3 RESULTS

Several anomalies on the windows were identified from photographic and video image data. A synopsis of the analyzed anomalies is provided in Table 8.2. Analyses of the temporal changes that occurred between different intervals of data collection (with time differences between 2 and 18 months) were affected by the limited information on window orientation, limited acquisition of consistent, high-resolution photography of the same windows across the different missions, and image registration problems.

Illustrated images of Base Block window anomalies (windows 1, 2, and 6) are provided in Figures 8.3 to 8.6. Enhanced images of Kvant-2 window V are shown in Figures 8.7 and 8.8, and Priroda window 1 is shown in Figures 8.9 and 8.10.

8.3.1 Base Block Windows

The Base Block module has been on orbit since 1986, and its windows had been exposed to the space environment for more than 12 years when surveyed. Because of their age, we anticipated contamination and MMOD strikes on the windows. We also documented significant changes on the windows occurring during the 18 months of this study. Specific information on our observations of the windows are presented below.

Base Block Window 6. On this window pane, two distinct anomalies were identified (Figure 8.3). The first anomalous feature is a cluster of small bright spots in the center of the outer window pane and the second anomalous feature is a probable MMOD impact area toward the outer edge on the outer pane. The cluster of bright spots in the center of the outer window pane is speculated to be caused by a probable MMOD impact that would have penetrated through the window cover and disintegrated into several fragments prior to the impact on the window surface. Based on this speculation and the dimensions of the cluster, it was estimated that a 2- to

TABLE 8.2 Description of the Window Anomalies Observed

Anomaly Category	Window	Dimensions	Estimated Figure
Probable MMOD impacts	Base Block window 1	~3–4 mm	8.6
	Base Block window 2	~2–7 mm	8.5
	Base Block window 6	~2–3 mm	8.3
	Kvant-2 window V	~5–6 mm	8.7
Translucent appearance	Kvant-2 window V	~40–60% area	8.7
	Priroda window 1	~75% area	8.9
Bright spots	Priroda window 1	~2–4 mm	8.10
	Kvant-2 window V	~3–5 mm (25)	8.7
	Kvant-2 window V	~1 mm (50)	8.7
Scratch/scuff marks	Base Block window 1	Across the window	8.6
	Base Block window 2	One quadrant	8.5
Unique patterns	Priroda window 1	Across the window	8.9
	Base Block window 6	Cluster of impacts	8.3

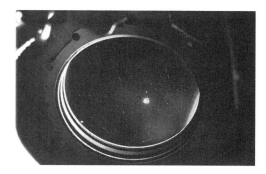

Figure 8.3 Overview image of Base Block window 6 shows a probable MMOD strike toward the left lower corner and a cluster of bright dots at the center of the window (35-mm film, NASA photograph NASA6-331-6).

3-mm-diameter hole could be found on the outer window cover. Efforts to obtain a high-resolution image of this Base Block window cover were not successful.

An additional set of bright spots were noted in the central region of the window surface from the NASA 7 video imagery (April 1998). Figure 8.4 shows comparative video image data from NASA 6 (December 1997) and NASA 7 (April 1998). The new bright spots could be recent deposits of contamination on the outer window pane.

Base Block Window 2. A new anomaly was identified from the NASA 7 video imagery that had not been seen before (see Figure 8.5). Imagery of this window

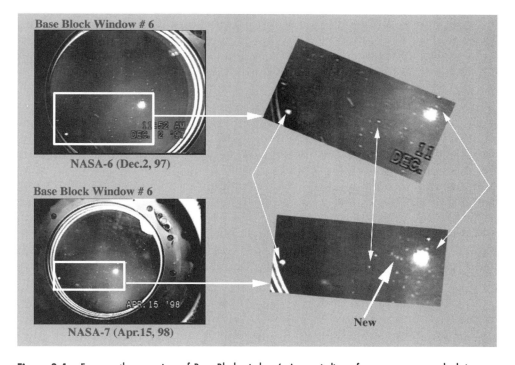

Figure 8.4 Four-month comparison of Base Block window 6. Arrows indicate features common to both images. Lower arrow marks a new set of bright spots, probably due to external contamination deposits (video, NASA6 ID 461707, NASA7 ID 461952).

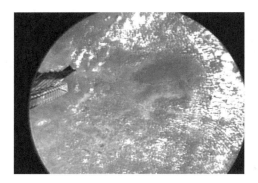 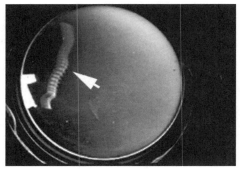

Figure 8.5 Images of Base Block window 2 show temporal changes between the left image (NASA 2, September 1996) and the right image (NASA 7, April 1998). On the right, the window shows a scratch/scuff mark toward the left of the image. The object on the left edge of each window image is the tip of a solar array panel (video, NASA2 ID 611787, NASA7 ID 461952).

obtained during NASA 2 (September 1996) provided a comparative view for analysis. The anomaly is a scuff mark that occupies about one-fourth of the visible window surface. It was probably caused by extravehicular activity (EVA) some time between NASA 2 and NASA 7, because the mark does not appear on the earlier NASA 2 image. Also, earlier analyses of the NASA 2 image identified two symmetrical dark spots (dimensions of 3 and 8 mm) and about five asymmetrically shaped impact areas (dimensions ranging between 2 and 7 mm) on the outer pane (not shown; see Image Science and Analysis Group, 1997).

Base Block Window 1. Base Block window 1 contains a large scratch mark that extends almost completely across the window and is similar to the window 2 scuff mark. This scratch mark was also caused by EVA-related activity (Figure 8.6). Also, toward the outer edge on the outer pane, a probable MMOD impact is visible. This MMOD impact is estimated be about 3 to 4 mm in size.

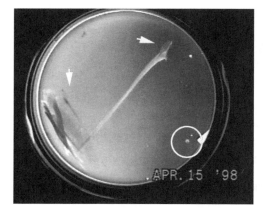

Figure 8.6 An overview image shows a scratch/scuff mark on Base Block window 1 (video, NASA7 ID 461952).

8.3.2 Kvant-2 Window V

The Kvant-2 module includes a large-diameter window (V) that was used for much of the Earth photography taken during the Shuttle–*Mir* program. The following anomalies were identified on this window pane: (1) existence of a translucent appearance around the periphery of the window, which was estimated to cover about 60% of the window surface (Figure 8.7 and 8.8); (2) about 50 small areas of bright spots of dimension less than 1 mm, that appear to spread throughout the surface of the window and were probably caused by contamination deposits; (3) about 25 large areas of bright spots of dimension ranging from 2 to 3 mm, probably caused by contamination deposits; and (4) a probable MMOD impact of 5 to 6 mm toward

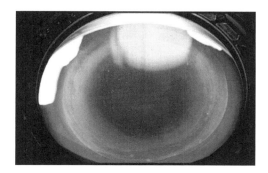

Figure 8.7 The prominent translucent appearance of the outer periphery of Kvant 2 window V covers nearly 60% of the window area, as observed during NASA 5 (35-mm film, NASA photograph NASA5-308-14).

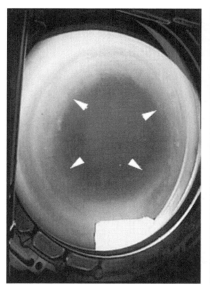
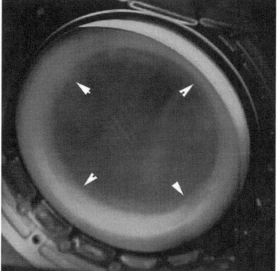

Figure 8.8 The haze/translucent area shown in video images of Kvant-2 window V decreased from 60% to 40% of the window surface area between July 1997 (left, video, NASA5 ID 461810) and February 1998 (right, video, NASA7 ID 461952).

the outer edge. Also, several bright spots were identified that appear to be arranged linearly on the exterior pane.

The translucency of this window was especially prominent around the outer edges during NASA 7. However, this translucent area appeared to be smaller on NASA 7 imagery (April 1998) than it was on NASA 5 imagery (July 1997; Figure 8.8). Because of the apparent difference between these two missions, the hazy appearance could be ephemeral and depend on *Mir* attitude and the lighting conditions.

8.3.3 Priroda Window 1

The Priroda module is the newest module on *Mir* and was launched into orbit in mid-1996. Because of their young age, the Priroda windows should represent windows in a more pristine condition.

A unique pattern that appeared like a symmetrical, rectangular shadow or an etched dark band in the center of the inner pane of the window was noticed during the window survey. Apparently, this pattern is not opaque and several Earth photographs were taken through this window. Astronaut A. Thomas (NASA 7) confirmed that this pattern is visible to the naked eye as well (personal communication). The origin of this pattern is unknown. Also, several concentric circles of translucent haze were noted (Figures 8.9 and 8.10).

From the video imagery, it was estimated that more than 50 bright spots, ranging from 2 to 4 mm in size, exist on the outer window pane. We interpret these bright spots as reflections from contamination deposits on the outer surface. Using video imagery obtained during the later part of NASA 7 (April 15, 1998) we identified an additional cluster of bright spots on the outer pane. This bright cluster of spots was not visible on earlier NASA 7 video imagery (February 12, 1998, Figure 8.10).

8.4 CONCLUSION

Our analysis of the *Mir* window survey shows that most defects were on the outer panes, resulting from external deposits of contamination and micrometeoroid impacts. Also, we have identified scuff marks on the outer panes which probably occurred during extravehicular activity. Video recording of the window panes pro-

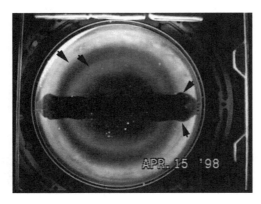

Figure 8.9 This overview image of Priroda window 1 in April 1998 shows a dark pattern in the center of the window and concentric rings of haze contributing to a translucent appearance away from the window center (video, NASA7 ID 461952).

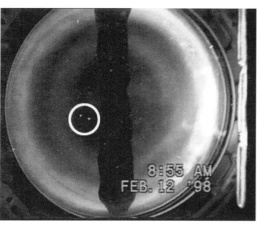 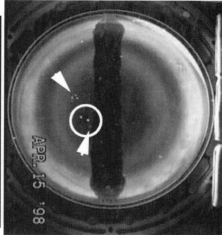

Figure 8.10 Temporal changes in Priroda window 1 occurred during NASA 7 and are illustrated in these video images (left, February 1998; right, April 1998; video, NASA7 ID 461952). Arrows point to new set of bright spots interpreted to be external contamination deposits.

vided many detailed views of thin scratches, deposits of contamination, smudges, and other anomalies on the external panes. Also, the video data had added value because of crew voice annotations.

The interior of the window housings were reported to be in pristine condition for all the windows surveyed, based on the analysis from the video imagery and crew's voice comments. Most interior window panes were also in pristine condition, with very few scratches and smudges. A NASA technical report (Image Science and Analysis Group, 1999) provides additional details on the window survey results.

ACKNOWLEDGMENTS

We sincerely appreciate the support and cooperation of our Russian collaborators S. Savachinco and I. Firsov of the RSC-Energia. We also express our sincere appreciation to NASA Johnson Space Center scientists G. Byrne, J. Disler, and C. Evans for their support and contributions to this investigation during the past three years.

REFERENCES

Image Science and Analysis Group. 1997 (March). *Mir Photo TV Survey (DTO-1118): STS-79 Mission Report.* Johnson Space Center Technical Document JSC-27761. Houston, TX: National Aeronautics and Space Administration.

Image Science and Analysis Group. 1999 (March). *Mir Window Survey: A Final Report.* Johnson Space Center Technical Document JSC-28578. Houston, TX: National Aeronautics and Space Administration.

The Caspian Sea

Chapter 9

Geographical, Geological, and Ecological Effects of Caspian Sea-Level Fluctuations: Introduction

Nikita F. Glazovskiy and V. A. Rudakov

Russian Academy of Sciences and Institute of Geography
Moscow State University
Moscow, Russia

ABSTRACT

This paper provides background material for the following chapters in this book. The Caspian Sea has experienced extreme sea-level fluctuations in the past, and has transgressed more than 2 m in the past 2 decades. The physical characteristics of the Caspian Sea (including water fluxes, salinity, regional climate) are discussed. The recent changes in sea level are introduced and are related to the regional environment and economy.

9.1 PHYSICAL CHARACTERISTICS OF THE CASPIAN SEA

The Caspian is the largest inland sea in the world (Figure 9.1), located in Eurasia between 47°07′ and 36°33′ north latitude and between 46°43′ and 54°50′ east longitude. The sea stretches more than 1200 km from north to south, and its width varies from 200 to 450 km. The length of its coastline is approximately 7000 km, the water surface area is approximately 380,000 km^2, and the volume of water is approximately 78 km^3. Those dimensions vary in response to significant short-term sea-level fluctuations. Average sea level over a period of many years has been lower than the world ocean level by 28 m.

Of the three segments of the Caspian—northern, central, and southern (Figures 9.2 and 9.3)—the northern Caspian is shallowest, with an average depth of approximately 5 m and an area of about 100,000 km^2; it contains only about 1% of the Caspian water volume. The northern and central Caspian subbasins are separated by the Mangyshlak rift. The central Caspian is a deep depression with a depth of up to 788 m. It has an area of 138,000 km^2 and a water-mass volume of 26,000 km^3. The central and southern Caspian segments are separated by the southern Apsheron rift,

Figure 9.1 Major political and geographical features of the Caspian Sea region.

the water depth above which does not exceed 180 m. The southern Caspian is a deep depression with a depth of 1025 m, the greatest of the entire sea. Its area is around 142,000 km^2, and its water-mass volume is over 50,000 km^3.

The Caspian Sea occupies a large depression (approximately 3.5 million square kilometers) of tectonic origin; mountain ranges define parts of the basin margin—the Caucasus, Talish, and Elburz to the Karatau (south-central Kazakhstan). A large part of the Caspian coast consists of the Pricaspian lowlands, western Turkmenistan, and Kura lowlands. The lowland elevation is −20 to −28 m, decreasing toward the Caspian; the Krasnovodsk and Mangyshlak plateaus reach higher elevations of +20 to +100 m. Mangyshlak has deep, dry, internally drained depressions of karst-deflation origin, which are as deep as 132 m (including some of the deepest inland depressions in the world).

Of the large bays around the Caspian, the most interesting and important is Kara-Bogaz-Gol. On the coast and offshore in the southern Caspian there are more than 100 active mud volcanoes; in the northern Caspian, more than 1600 salt domes have been found, in some of which salt is at the ground surface.

9.1.1 Climate

Most of the Caspian Sea and coastal regions have a moderate or harsh continental climate—summer air temperatures reach +40°C on the western shore and +47°C on the east. High temperatures result in intense evaporation on the one hand, but the significant diurnal temperature differential helps promote moisture condensation in the coastal soil at night. In the winter the air temperature is +5 to −10°C along most of the coastline and lower in the mountains.

Figure 9.2 Regional view of the northern Caspian Sea, April 1991 (NASA photograph STS039-151-007). This south-southeastward view was taken midway between the onset of transgression and the present; it spans the northern Caspian, the Volga and Ural River deltas, and the Apsheron and Mangyshlak Peninsulas. The Aral' Sea is visible in the eastern distance, but most of the Gulf of Kara-Bogaz-Gol is beneath clouds.

Figure 9.3 Regional view of the southern and eastern Caspian Sea, October 1997 (NASA photograph NASA6-707-031). In this recent photo of the southeastern Caspian, Kara-Bogaz-Gol is full; in sunglint, the channel that was dredged through the spit separating the bay from the Caspian can be seen. The channel was partially opened in 1984, then cut through in 1992 to allow rising Caspian waters to spread over the bay floor, where there was little water in 1985. The wing-shaped Cheleken Peninsula is near the center of view and Gorgan Bay (Iran) occupies the southeast corner of the Caspian.

Total annual precipitation above the sea and for most of the coast is 100 to 400 mm, increasing to 700 to 800 mm in mountainous areas. Maximum precipitation occurs in the Lenkoran region, where it reaches 1100 mm under subtropical climatic conditions. Because of the high temperatures and dry air, evaporation exceeds precipitation by 1.5- to 3-fold. Only in Lenkoran and mountainous areas does precipitation exceed evaporation.

In a number of coastal areas, desert formation is increasing. In Kalmykia, the first European anthropogenic desert has formed, and the sands there are already encroaching on the capital of Elista. In the Astrakhan region, desert formation is especially intense in five administrative areas; northern Dagestan is also dealing with this problem.

The Caspian is characterized by strong winds: the average annual wind speed is about 6 m/s. Strong winds make the water significantly choppier and carry salt from the sea onto dry land.

9.1.2 Salinity

The salinity of the Caspian Sea increases from 1‰ in the north to 14‰ in the south. Average salinity is 12.9‰ predominated by magnesium sulfate salts. The disparity between the Caspian Sea salinity and that of the ocean is tied to the paleogeographic history of the basin and certain characteristics of its salt content formation.

Water circulation of the Caspian is mainly counterclockwise. Intensive water exchange among the northern, central, and southern parts of the sea is important because the waters, which are enriched with oxygen and cooled in the winter in the shallow areas of the northern Caspian, flow into the depression of the central Caspian and then join the southern Caspian. Thus, deep parts of the sea are aerated.

9.1.3 Water Balance

The main incoming parameters of water balance are river flow and atmospheric precipitation; the outgoing parameters are evaporation and the flow of water into the Gulf of Kara-Bogaz-Gol (Table 9.1). The largest river feeding the Caspian is the Volga, which drains about 20% of the area of Europe; more than 80% of the river waters in that drainage basin enter the sea. Forty-two percent of the population of Russia lives around the Volga Basin, and more than half of Russia's industrial development is there. Human activities have significantly transformed Volga River flow and have led to pollution. Parameters in the water balance such as evaporation and groundwater flow have not been studied thoroughly. Most parameters are subject to significant fluctuations that, in turn, result in changes in Caspian water balance.

9.2 ECONOMY OF THE REGION

The coast and water area of the Caspian Sea belong to five countries: Russia, Kazakhstan, Turkmenistan, Iran, and Azerbaijan. A number of large coastal cities are within this economically important region: Baku, Lenkoran, Bandar-e Anzali, Cheleken, Turkmenbashi (previously Krasnovodsk), Bekdash, Aktau, Astrakhan, Makhachkala, and Derbent.

Great biological diversity is an inherent trait of the Caspian and its littoral zone: Approximately 400 animal species are indigenous. A number of areas (in the Volga delta, Krasnovodsk Bay, and on the coast of Azerbaijan) are very important places of habitation and molting of waterfowl. The Caspian has an ancient tradition of fishing. Its basin accounts for more than 80% of the world catch of sturgeon and approximately 90% of the black caviar. Herring, perch, sprat, and seals are also caught.

TABLE 9.1 Average Water Balance for the Caspian Sea Before the Last Sea-Level Rise

Incoming (km^3/yr)		Outgoing (km^3/yr)	
Atmospheric precipitation	71.6	Evaporation	374.5
Surface flow	286.7	Flow into Gulf of Kara-Bogaz-Gol	11.6
Groundwater flow	2.0	Diversion of water	0.8
Total gain	360.3	Total loss	387.1

Data from Remizova (1969).

Enormous oil and gas reserves have been found beneath the Caspian and along the coast, and there are ongoing disagreements among several countries regarding jurisdiction over Caspian subsea petroleum resources. According to existing data, oil reserves within the Caspian region in Azerbaijan are estimated at 500 million tons, while those in Russia are estimated at 600 million tons. The largest reserves were discovered in Kazakhstan and Turkmenistan. Although oil fields have been developed in the Caspian since the beginning of the twentieth century, the discovery of new oil and gas fields has made the Caspian region one of the most important potential oil and gas regions in the world. At present, the majority of Caspian region countries have created consortia to build large new pipelines from the Caspian to the Black Sea. Discussions have begun concerning the possibility of building such pipelines from the Caspian to the east and south to deliver oil and, especially, gas to eastern and southern Asia.

In the Gulf of Kara-Bogaz-Gol, sulfates are recovered, while on the Cheleken Peninsula, a field is being developed for production of iodine and bromine brines. On the Mangyshlak (Kazakhstan) Peninsula, there is a uranium field that provides fuel for an atomic power plants.

Agriculture is one of the most important aspects of the economy. Along most of the coastline there is pasture that supports ranching. On the western shore of the Caspian, vineyards and gardens cover a significant area. Melons and gourds are cultivated everywhere; Astrakhan watermelons are particularly popular. Cotton is grown on the irrigated lands of Azerbaijan and Turkmenistan; subtropical crops are cultivated in the regions of the Lenkoran and the Atrek deltas.

Shipping is common on the Caspian Sea; rivers and channels link the Caspian to the Black, Azov, Baltic, and White Seas. A railroad ferry operates between Baku and Turkmenbashi.

Before the fall of the Soviet Union, sanatoriums, boarding houses, and vacation resorts in Azerbaijan and Dagestan attracted many vacationers to the western coast of the Caspian. Today, the coast is not as widely used for recreational purposes because of economic problems.

9.3 SEA-LEVEL FLUCTUATIONS AND THEIR IMPACT

Because the Caspian is landlocked, sea level is very sensitive to changes in the water-balance parameters and, according to some researchers, to tectonic changes. During the last 20,000 years (i.e., when human settlements already existed on the coast) following separation of the Caspian from the Azov/Black Sea basin, Caspian sea level changed by 100 m. In historical times, Caspian sea level has also changed by 5 to 6 m. Over the course of almost a century (1837–1933) there was comparatively insignificant fluctuation: in the range −25.3 to −26.5 m. From 1933 to 1940, the sea level dropped sharply, by 1.7 m. Afterward, the level continued to fall slowly (although despite the general decrease, there were small rises) until 1977, when it reached the lowest point ever recorded by scientific instruments: −29 m.

However, since 1978, the Caspian has been rising sharply again. In the past 20 years, sea level has risen by 2.6 m, reaching −26.9 m. Between 1978 and 1993, the average increase was 13 to 14 cm/yr. In 1994, Caspian sea level rose by a dramatic 26 cm. [*Editors' note:* Some of the changes between 1982 and 1997 are illustrated in

Chapter 10.] Although there are various explanations for the periodic fluctuations in level, the actual causes have not been determined, and there is no predominant opinion regarding subsequent forecasts of fluctuations. Nonetheless, most forecasts assume that sea level will continue to rise until sometime between 2005 and 2010.

The fall of sea level over 44 years favored certain ways of developing the Caspian coastal zone. The present economic potential of the Caspian coast was realized during the regression of −27 to −28 m. At that time, the coastal zone was settled rapidly, giving rise to a series of populated areas and infrastructures. Between 1978 and 1995 the area of the sea increased by 11.2%, which led to variable coastal flooding of settlements as well as industrial facilities. This was the period of active growth of industries, agriculture, and land development, including development of the Caspian Sea zone. On the shores, numerous settlements, sanatoriums, and vacation resorts were built. Transportation and power-supply networks were created. In a number of coastal areas, oil recovery began.

The rapid increase in sea level that began in 1978 soon led to complex environmental and socioeconomic problems. On the shore of the northern Caspian, the situation was aggravated by short-term but quite frequent increases in sea level, during which hundreds of square kilometers of the coast were flooded. Threatened by flooding and destruction are such large and economically important cities as Makhachkala, Derbent, and Kaspiysk (Dagestan); Lagan (Kalmykia); Kamyzyak (Astrakhan); Baku (Azerbaijan); and Turkmenbashi and Bekdash (Turkmenistan).

According to present estimates, if sea level reaches −25 m, the total damage to the economy in the coastal areas may have a significant impact on the regional economy. Therefore, the government of the Russian Federation has issued a series of laws and has instituted a program to reduce and prevent the negative consequences of Caspian sea-level increases. The total cost of measures to protect the population and prevent flooding of economic projects and other facilities on the Russian sector of the Caspian shore was $60 million in 1996.

Unfortunately, at present, the complex economic situation in Russia has allowed for implementation of only 17% of the planned measures. As it is unlikely that there will be a significant and rapid improvement in the economic situation in the near future, complete realization of the program is doubtful. In this regard it is important to define more important and effective measures to reduce and prevent damage—measures that could be implemented using existing finances. Currently, there is no widely accepted forecast regarding future changes in Caspian sea level. Therefore, in our opinion, any strategy for economic activity should be flexible enough to adapt to changing levels rather than struggle to overcome the consequences of periodic rises and falls.

Finally, it is important to take into account prospects for development of various industries in the region. Hydrocarbon reserves have been located in the coastal zones of all the members of the Russian Federation and in other coastal states. Therefore, it is extremely important to consider changes in sea level when developing these natural resources.

The sea-level rise is forcing population migration, considering the forecasted increase up to −25 m before the year 2005. According to one pessimistic scenario, if protective measures are not taken in the Astrakhan, Kalmykia, and Dagestan regions, eight cities and small towns and 105 rural settlements with a total population of about 197,000 people will be within a zone negatively affected by the Caspian Sea. One particular danger is flooding of populated city areas, where most of the coastal dwellers are concentrated, along with recreation facilities and industrial and nonindustrial infrastructures. A similar dilemma exists in other coastal states. In Kazakhstan, 75 settled localities with a population of around 305,000 people could be flooded.

9.4 OTHER ENVIRONMENTAL PROBLEMS

On the Caspian coast large industrial facilities are seriously polluting the environment—for example, the gas-condensate facility in the Astrakhan region. Other problems there raise the possibility of accidental release of extremely toxic gases. On the Cheleken Peninsula (Turkmenistan), water polluted during the production of iodine and bromine brines is dumped into the sea. In Sumqayit (Azerbaijan), water used in chemical production is also disposed of in the Caspian. Pollution can be observed in virtually all river basins and in the sea basin itself.

Disposal of highly mineralized groundwater extracted during oil recovery, and oil spills, have led to the formation of special lakes in areas of oil production in Kazakhstan on the northeastern coast of the Caspian. They pose a serious threat to ecosystems inasmuch as catastrophic levels of sea pollution could occur during wind-driven high tides or in the case of a further increase in the sea level.

9.5 RESULTS OF INVESTIGATIONS

The goal of monitoring nature and the economy in the Caspian region using space photographs has been to assess changes in the natural and socioeconomic conditions of the coast as well as the sea resulting from anthropogenic effects and the rising of the sea level. These data are necessary in order to formulate recommendations for strategies of natural resource use. Results of investigations of major geoecological problems are reported in nine articles by Russian authors (Figure 9.4).

The need for further study of the geomorphological and geological structures of the northern Caspian coastline is emphasized by Aristarkhova et al. in Chapter 11. Interpreting ground surface changes from images taken from spacecraft, including some from Space Shuttle, is one way to obtain additional information on the deep structure of the territory and on the sedimentary cover.

The migration of the delta of the Volga, Europe's largest river and the main source feeding the Caspian Sea, is analyzed in Chapter 12 by Alekseevskiy et al. In recent years, shallow streams have been disappearing and water flow has been more concentrated in the largest channels. This is tied to the fact that river flow is regulated by several large reservoirs. A rise in sea level poses a threat to engineering structures, the roadway system, and all coastal infrastructures.

Changes in bird habitats are closely tied to sea-level dynamics. The marshlands of the Volga delta are a key part of the migration paths of hundreds of avian species traveling from the subtropics to Siberia and the European part of Russia. The Volga delta is the largest delta in Europe and is home to a biosphere preserve. In Chapter 13 Baldina et al. examine changes in bird habitats and analyze the interaction of the Caspian sea level and Volga delta. Additional and continuous photography of the delta from space is required for continued habitat monitoring.

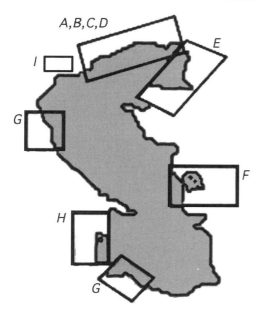

Figure 9.4 Geographic coverage of Chapters 11 to 19. Proceeding clockwise from the Volga River delta: A, Chapter 11; B, Chapter 12; C, Chapter 13; D, Chapter 14; E, Chapter 15; F, Chapter 16; G, Chapter 17; H, Chapter 18; I, Chapter 19.

Changes in coastal vegetation, one of the most dynamic components of nature, are discussed by Kravtsova and Myalo in Chapter 14. Space photos allow scientists to track changes in relief and in vegetative cover of the coastal zone as sea level rises. Coastal zone dynamics of the northeastern Caspian are examined by Kravtsova in Chapter 15. This part of the sea is shallow: In a coastal strip of 50 km, the depth does not exceed 3 m; therefore, even an insignificant oscillation in sea level strongly influences the vast coastal zone. In recent years the territory has been subjected to extensive partial flooding and saturation with the brackish water of the depressions. Seafloor topography that formed during the regression has also been modified.

The Gulf of Kara-Bogaz-Gol in the eastern Caspian is a unique region. Its historical evolution attests to the strong impact of natural factors and to interference by humankind—namely, isolation of the bay from the sea by construction of a dam and the subsequent removal of the barrier. In Chapter 16 Varuschenko et al. look at the history of the bay and the changes in the subaqueous area up to the time when almost all the water had vanished. The dynamics of the coastal zone are shown in the transgressive and regressive periods of Caspian Sea state. Destruction of the dam may allow for renewal of the bay and preservation of its natural landscape and unique saline resources.

Shipilova (Chapter 17) uses space photographs to analyze the movements of water masses along the Caspian coast. As the sea level changes, bottom topography and coastline morphology affect the movements of nearshore waters. In Chapter 18 Ignatov and Solovieva discuss the most dynamically active zones of the western Caspian coast and the influence of sea level rise on coastal morphology. Shestakov (Chapter 19) takes a look at structural changes in land use in the Caspian coastal zone that have resulted from ground destruction caused by full and partial flooding. Space photographs have been used to record ground features, ascertain the extent of flooding, and forecast the expected level of damage.

9.6 CONCLUSION

The Caspian Sea is a unique basin on our planet—the largest landlocked lake in the world. The Caspian coastal zone comprises many different landscapes. On the sea and coast, there are natural structures such as the Gulf of Kara-Bogaz-Gol, large inland depressions, mud volcanoes, and salt domes. The Caspian is a region rich in natural resources, especially biological (e.g., the world's largest shoal of sturgeon) and hydrocarbons. Although the Caspian Sea and its coastline were already considered an important region for oil recovery, new oil and gas deposits discovered in recent years have increased the significance of the region. Oil and gas field development has attracted powerful investors from all over the world, and intensifying activity demands close attention to protection of the environment.

Fluctuations in Caspian sea level over many years have led to a number of complex ecological, social, and economic problems. In all the studies cited above, theories on the course of development of natural processes were evaluated in an environment being modified by outside factors. This is of global significance because as the world ocean rises, the Caspian can be used as a model to predict changes that will occur in coastal zones worldwide and to study the stability of global coastal systems. The Caspian Sea is also a model showing possible socioeconomic processes that will be observed if the world's oceans rise under conditions of global warming.

The most important goal of the Russian–U.S. program for monitoring the Caspian region from space is to assess changes in the natural and socioeconomic conditions of the coastal zone and the sea resulting from both human activities and rising sea level. These data will contribute to recommendations for strategies of natural resource protection and development. These cooperative studies should help resolve issues such as the contribution of modern tectonism to sea-level change, and the extent and dynamics of oil-related pollution, landscape transformation, and land use. Existing photographs spanning 30 years for many parts of the world provide a foundation for future studies. Significant observations (both long and short term) from the International Space Station will facilitate future geoecological monitoring of the Caspian region.

ACKNOWLEDGMENTS

The authors gratefully acknowledge the assistance of R. Allen (Dickinson College) and J. A. Robinson in preparing the maps. Citations of recent papers in English on the Caspian Sea were compiled by J. A. Robinson. We also thank TechTrans International, Inc. (Houston) for translating the paper into English and C. A. Evans and J. Caruana for constructive reviews.

REFERENCES

Remizova, S. S. 1969. Water balance. In *Caspian Sea*. Moscow: Moscow State University, pp. 107–138.

Editor's Note: English-speaking readers will find the following list of recent literature on the Caspian Sea to be a useful starting point for further information.

Anonymous. 1994. The rising Caspian Sea. *The Economist,* 331(7860):52.
Anonymous. 1998. Caspian Sea geology [this multipart series on the Caspian Sea outlines source history, regional structure, and petroleum potential]. *Offshore,* 58:75–77.
Bortnik, V. N., Kuksa, V. I., and Saltankin, V. P. 1997. The present-day geoenvironmental situation in the basins of the Volga and the Caspian Sea. *Water Resources,* 24:506–512.
Dumont, H. 1995. Ecocide in the Caspian Sea. *Nature,* 377:673.
Efendieva, I. M., and Dzhafarov, F. M. 1993. Ecological problems of the Caspian Sea. *Hydrotechnical Construction,* 27:13–14.
Finlayson, C. M., Volz, J., and Chuikov, Y. S. 1993. Ecological change in the wetlands of the lower Volga, Russia. In Moser, M., Prentice, R. C., and Van Nessem, J., eds., *Waterfowl and Wetland Conservation in the 1990s: A Global Perspective.* IWRB Special Publication 26, Slimbridge, Gloucester, England: International Waterfowl and Wetlands Research Bureau, pp. 61–66.
Glantz, M. H., and Zonn, I. S., eds. 1997. *Scientific, Environmental, and Political Issues in the Circum-Caspian Region: Proceedings, NATO Advanced Research Workshop on the Scientific, Environmental, and Political Issues of the Circum-Caspian Region,* Moscow, May 13–16, 1996. NATO ASI Series, Partnership Subseries 2, Environment, Vol. 29. Boston: Kluwer Academic, 312 pp.
Golitsyn, G. S., Ratkovich, D. Ya., Fortus, M. I., and Frolov, A. V. 1998. On the present-day rise in the Caspian sea level. *Water Resources,* 25:117–122.
Ignatov, Ye. I., Kaplin, P. A., Lukyanova, S. A., and Solovieva, G. D. 1993. Evolution of the Caspian Sea coasts under conditions of sea-level rise: model for coastal change under increasing "greenhouse effect." *Journal of Coastal Research,* 9:104–111.
Kaplin, P. A., and Selivanov, A. O. 1995. Recent coastal evolution of the Caspian sea as a natural model for coastal responses to the possible acceleration of global sea-level rise. *Marine Geology,* 124:161–176.
Khodorevskaya, R. P., and Novikova, A. S. 1995. Status of beluga sturgeon, *Huso huso,* in the Caspian Sea. *Journal of Ichthyology,* 35:59–71
Kupriyanova, Ye. I., and Trapeznikova, O. N. 1998. Changes in natural and economic conditions in Kalmykia associated with the rising level of the Caspian Sea. *Mapping Sciences and Remote Sensing,* 35:193–201.
Mamedov, A. V. 1997. The Late Pleistocene–Holocene history of the Caspian Sea. *Quaternary International,* 41/42:161–166.
Meleshko, V. P., Golitsyn, G. S., Volodin, E. M., Galin, V. Ya., Govorkova, V. A., Meshcherskaya, A. V., Mokhov, I. I., Pavlova, T. V., and Sporyshev, P. V. 1998. Calculation of water balance components over the Caspian Sea watershed with a set of atmospheric general circulation models. *Izvestiya (Atmospheric and Oceanic Physics),* 34:534–542.
Meshcherskaya, A. V., and Aleksandrova, N. A. 1993. Caspian sea level forecast from meteorological data. *Russian Meteorology and Hydrology,* 3:52–60.
Meshcherskaya, A. V., Aleksandrova, N. A., and Golod, M. P. 1994. Temperature and precipitation regime in the Volga and Ural basins and an estimation of its impact on the Caspian sea level fluctuations. *Water Resources,* 21:427–435.
Meshkani, M. R., and Meshkani, A. 1997. Stochastic modelling of the Caspian sea level fluctuations. *Theoretical and Applied Climatology,* 58:189–196.

Nikolenko, A. V. 1997. On the long-period variations in the Caspian sea level. *Water Resources*, 24:235–239.

Novikov, Yu. V. 1987. Monitoring human impacts on the Kara-Bogaz-Gol. *Mapping Sciences and Remote Sensing*, 24:123–128.

Ratkovich, D. Y. 1997. Water balance of the Caspian Sea and its level regime. *Hydrotechnical Construction*, 31:355–363.

Remizova, S. S., and Myagkov, M. S. 1995. On the problem of long-term forecasting the Caspian sea level. *Water Resources*, 22:311–317.

Rodionov, S. N. 1994. *Global and Regional Climate Interaction: The Caspian Sea Experience.* Water Science and Technology Library, Vol. 11. Boston: Kluwer Academic, 241 pp.

Rychagov, G. I. 1997. Holocene oscillations of the Caspian Sea, and forecasts based on palaeogeographical reconstructions. *Quaternary International*, 41/42:167–180.

Sedletskii, V. S., and Baikov, A. A. 1997. The nature of Caspian sea level fluctuations. *Lithology and Mineral Resources*, 32:208–217.

Sidorenkov, N. S., and Shveikina, V. I. 1996. Changes in the climatic regime of the basins of the Volga and the Caspian Sea in twentieth century. *Water Resources*, 23:369–374.

Svitoch, A. A., and Yanina, T. A. 1996. "Cold" and "warm" transgressions of the Caspian Sea. *Oceanology*, 36:277–281.

Tarasov, A. G. 1996. Biological consequences of pollution of the Caspian Sea basin (prior to 1917). *Water Resources*, 23:416–425.

Uibopuu, H.-J. 1995. The Caspian Sea: a tangle of legal problems. *The World Today*, 51:119–123.

Vakulovsky, S. M., and Chumichev, V. B. 1998. Radioactive contamination of the Caspian Sea. *Radiation Protection Dosimetry*, 75:61–64.

Vaziri, M. 1997. Predicting Caspian Sea surface water level by ANN and ARIMA models. *Journal of Waterway, Port, Coastal, and Ocean Engineering*, 123:158–162.

Winkels, H. J., Polman-Gorlova, T. A., and Lukaikin, V. 1996. *Selection of Presentations, Russian–Dutch Scientific and Technical Workshop, Research for Sustainable Development of the Volga Delta in View of Caspian Sea Level Rise* (in English and Russian), Astrakhan, Russia, April 10–12. Lelystad, The Netherlands: Ministry of Transport, Public Works and Water Management, Institute of Inland Water Management and Waste Water Treatment, 124 pp.

Zakharova, E. A., and Savenko, V. S. 1998. Fluoride and boron of the Volga River and the Caspian Sea waters mixing zone. *Geochemistry International*, 36:180–182.

Zektser, I. S., Plemenov, V. A., and Kas'yanova, N. A. 1994. Role of recent tectonics and mud volcanic activity in the water and salt balance of the Caspian Sea. *Water Resources*, 21:404–409.

Zubakov, V. A. 1993. The Caspian sea level oscillations in the geological past and its forecast. *Russian Meteorology and Hydrology*, 8:65–70.

Chapter

10

A Caspian Chronicle: Sea-Level Fluctuations Between 1982 and 1997

Patricia Wood Dickerson

Office of Earth Sciences
NASA Johnsoxn Space Center
Houston, Texas USA

ABSTRACT

In the present episode of sea-level change, waters of the Caspian (Figure 10.1) regressed to their lowest level in 1977; by 1985 there were clear indications that transgression was under way, and over the past 20 years sea level has risen 2.6 m.

Figure 10.1 Shaded relief map of Caspian region. Locations numbered 2 to 7 correspond to Figures 10.2 to 10.7. (Modified from U.S. Geological Survey, 1996.)

The effects of relative fall and rise have been most visible along the northern shore and in Gulf of Kara-Bogaz-Gol, where topographic relief is subdued. By the time of the Shuttle–*Mir* missions, however, effects of rising sea level could also be seen in areas of steeper topography, such as Gorgan Bay on the southeastern coast and around Apsheron Peninsula. The photographs of this series (Figures 10.2 to 10.7, and 19.2, see color insert for Figs. 10.3, 10.5, and 19.2) are in clockwise progression from the Volga River delta in the northwest to Apsheron Peninsula in the southwest.

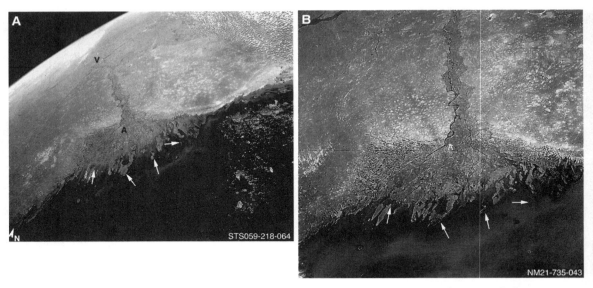

Figure 10.2 Volga River delta. The Volga is the largest river in Europe and drains an area of 1,380,000 km². (*A*) April 1994 (NASA photograph STS059-218-064). Astrakhan (A on the southwest side of the delta) and Volgograd (V at the right-angle bend in the river) appear as gray patches in this northwestward view. (*B*) May 1996 (NASA photograph NM21-035-043). In the two years between these two photographs, the islands south of the delta have become submerged. Interdune areas along the delta margins appear wetter as well. Several areas of change have been marked with arrows to facilitate comparison. See also Fig. 19.2 in color insert.

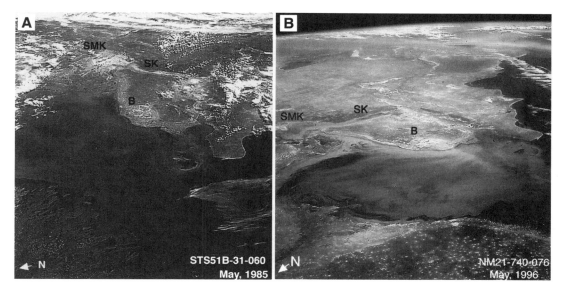

Figure 10.4   Buzachi and Mangyshlak Peninsulas: NASA photographs (*A*) May 1985 (STS51B-031-060); (*B*) May 1996 (NM21-740-076). Between these two southeastward photographs, taken seven years apart, the most striking difference is in Sor Kaydak (SK) and Sor Mertvyy Kultuk (SMK); the salt flats (*sor*) stretching to the east and southeast of the peninsula were dry in 1985 and inundated in 1996. On Buzachi Peninsula (B) much of the western shore has been flooded.

Figure 10.6   Gorgan Bay, Iran. (*A*) April 1982 (NASA photograph STS003-010-586). Taken five years after the 1977 lowstand, this southwestward view of the fault-controlled southeastern Caspian coast portrays a narrow inlet to Gorgan Bay with a squat promontory just north of the channel. Note the western limit of Gorgan Bay waters, as well as the position of the Iran–Turkmenistan shoreline extending northward from the inlet. (*B*) November 1995 (NASA photograph STS074-708-035). By 1995 bay waters had transgressed westward, parallel to the Elburz Mountains, increasing the area of the bay by roughly one-half; the spit that defines the northern bayshore is longer and narrower, owing to the westward flooding (arrow). Waters also encroached upon the eastern shore, widening the channel into the bay and changing the shape of the promontory. A lagoon, with a narrow outer bar (arrow), now covers what had been beach.

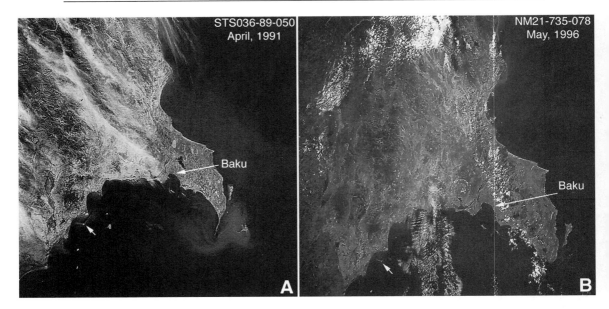

Figure 10.7 Apsheron Peninsula, Azerbaijan: NASA photographs (*A*) April 1991 (STS036-089-050); (*B*) May 1996 (NM21-735-078). Rising waters are a concern for petroleum production facilities on- and offshore around the Apsheron Peninsula. Subtle changes due to rising sea level can be seen along this coastline, where topographic relief is higher, by comparing the two views (arrows).

ACKNOWLEDGMENTS

The author gratefully acknowledges the assistance of R. Allen (Dickinson College) in locating and processing images and in preparing the index map. Constructive reviews by J. Caruana, C. A. Evans, and J. A. Robinson are also appreciated.

REFERENCES

St. George, G. 1974. *Soviet Deserts and Mountains*. Amsterdam: Time-Life Books, p. 29.
U.S. Geological Survey. 1996. GTOPO30: Global 30 arc second elevation data set. *http://edcwww.cr.usgs.gov/landdaac/gtopo30/gtopo30.html*

Chapter

11

Morphological and Geological Structure of the Northern Coast of the Caspian Sea

L. B. Aristarkhova, A. A. Svitoch, and O. N. Bratanova

Department of Geography
Moscow State University
Moscow, Russia

ABSTRACT

Photographs taken from *Kosmos, Mir,* and Space Shuttle spacecraft over the past 20 years permit integration of structural, geomorphic, and sedimentologic data from the northern Caspian basin. Linear features and other landforms observed on space photos, such as those of the Ural River delta and Dossor area, were verified by means of geophysical data and field observations. Such local structures could then be related to the broader tectonic framework of the region. Pliocene to Quaternary fluvial, deltaic, and coastal plain lithofacies of the Volga River valley have been distinguished on the basis of tonal and textural differences seen in Space Shuttle photographs and confirmed by outcrop examination.

11.1 INTRODUCTION

A coastal shelf in the recent past, the northern coast of the Caspian Sea is located south of the major oil and gas basin of the Caspian depression. Many geological features and structures require further study, including surficial deposits, salt-cored and subsalt structures in the sedimentary succession, and deep-seated structures. Geological and geomorphological interpretation of images of Earth's surface taken from spacecraft is one way to obtain additional insight about these structural features.

11.2 RESULTS

11.2.1 Deep-Seated Structures

Structural and geomorphological interpretation of space photographs makes it possible to use surface features to discern structural shapes lying at varying depths beneath unconsolidated surficial deposits, along with the most recent trends of tectonic activity. The specific objective of morphostructural interpretation of orbital photographs is to identify morphological lineaments and structures.

Morphological lineaments are manifested on photographs by straight-line or arc patterns that represent topographic and drainage features. On the northern coast of the Caspian, these include shallow ledges, slope discontinuities, sharp bends, and/or linear segments in streams, lake chains, solonchaks (salt pans), aeolian forms, configurations of coastlines, and abrupt changes in relief. As a rule, lineaments indicate the trends of fractures and faults activated in the most recent tectonic phases.

Morphological structures are areas of Earth's surface where geomorphic features differ from those of adjacent regions. Morphological structures may be formed by tectonic processes and modified by other forces that shape topography. In orbital photographs, morphological structures can be identified (1) by a characteristic topographic pattern similar to the image of the tectonic structure, and (2) by geomorphological and landscape "anomalies" differing in structure, photographic hue, and color from the regional background.

For the Caspian, the geological structure has long been studied by interpretation of aerial photographs and mosaics, and recently of space photographs of varying scale and type. Here we interpret space photographs for morphostructural purposes as part of independent investigations and during analyses of topographic maps (Aristarkhova and Fedorova, 1982). We present a description of the morphological structures for several regions of the Caspian and data concerning the most recent tectonics and deep-seated structures, including faults, which were largely confirmed by data from recent geological and geophysical studies (Slepakova, 1977; Aristarkhova et al., 1991).

Figure 11.1 (*Kosmos* satellite photograph Ф-2170-К 426, 1978) depicts regional morphological lineaments running north-northwest and north-northeast (lineaments 1 and 2), which coincide with Ural River delta channels, several less extensive lineaments, and several structures related to salt dome tectonics and the large Dossor structure (D). Comparison of the photointerpretation with geological and geophysical data produced four major results. First, the north-northwest regional morphological lineament is part of a mobile zone of deep-seated faults extending south of the Russian platform from the Pachelm depression in the northwest to Ustyurta to the southeast (Slepakova, 1977; Aristarkhova, 1981). Second, the north-northeast regional morphological lineament is a long-lived mobile zone that extends into the underlying salt deposits (Aristarkhova et al., 1991; Aristarkhova and Pryakhina, 1994). Third, the small morphological structures appear as salt domes and a salt-withdrawal syncline filled by Lake Zhaltyr. Fourth, the large oval Dossor structure east of the Ural River delta corresponds to the surface expression of the basement block and the subsalt brachyanticline (Aristarkhova and Pryakhina, 1994).

Earlier data were supplemented by color space photographs of the northern coast of the Caspian taken in 1996 during joint Shuttle–*Mir* flights (see also Figures 10.2

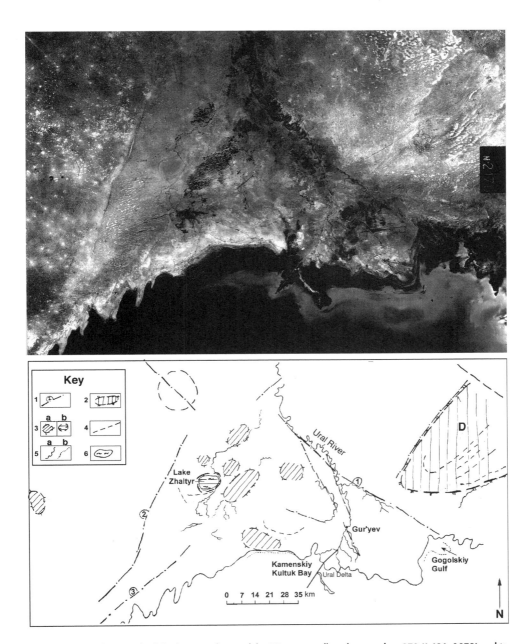

Figure 11.1 Photograph of the lower Ural River delta (*Kosmos* satellite photograph φ-270-K 426, 1978) and its geomorphological interpretation. Symbols for interpretation (Figures 11.1 to 11.5): 1, morphological lineaments labeled by number; 2, large positive morphological structures associated with the most recent activation of base blocks and subsalt arched uplifts: Astrakhan (A), Dossor (D); 3, local morphological structures caused by salt tectonics [3a, positive structures (salt domes); 3b, negative structures associated with active compensation synclines, specifically the Inder and Zhaltyr lakes]; 4, structural lines that pinpoint the location of various structural tectonic elements; 5a, waterways and modern coastlines; 5b, flooded coastlines; 6, lakes.

and 10.3). In the Ural River delta region (Figure 11.2), the central part of the Dossor structure is outlined and morphological lineaments that complicate the structure are marked. The Volga River region clearly exhibits west-northwest-trending regional morphological lineaments that intersect the river delta (lineaments 2a and 3, Figure 11.3), along with an extensive east-northeast-oriented morphological lineament (4) that intersects the Volga valley at the Zamyana settlement. The link between this morphological lineament and deep faults was traced on site and confirmed using geophysical data. Several other linear and arcuate morphological lineaments correspond directly to deep-seated structures. The largest of them is the Astrakhan (A) structure, surrounded by arcuate morphological lineaments and intersected by a northeast-trending regional morphological linear element (5) that crosses the subsalt Astrakhan crest and associated Astrakhan gas condensate field.

The coast of the Caspian Sea from the Volga River delta to the Ural River delta was photographed from the Space Shuttle (STS059-218-064; Figure 11.4 see also Figure 19.2 color insert), and regional morphological lineaments 2a and 3 are distinctly visible. Bordering the Ural River delta from the west, lineament 2 merges into lineament 2a. It is arcuate and open to the north. Apparently, the system of morphological lineaments 2, 2a, and 3 partially frames a huge annular morphological structure. The photograph also exhibits regional morphological lineament 6 extending from the Khaki solonchak to the south-southeast. It pinpoints the deep-seated Azgir fault interpreted previously from geological and geophysical data (Zhuravlev and Kuzmin, 1960). The strike of this fault and its length are well established.

We compiled data from the photointerpretation presented in Figures 11.1 to 11.4 to a common base map (Figure 11.5); the data sets correlate well and complement each other. Geological and geophysical data confirm a link between the morphological lineaments and structures interpreted from photos and the corresponding deep-seated tectonic structures. Morphostructural interpretation of photographs is particularly effective for identifying deep-seated fault and fracture zones that have undergone Quaternary reactivation. The large positive Dossor- and Astrakhan-type structures associated with movement of subsalt uplifts and, evidently, of basement blocks are also clearly visible.

In the northern Caspian, orbital photographs are less useful for identifying small local morphological structures caused by salt tectonics. Only the most active of the local morphological structures associated with salt tectonics can be interpreted reliably. These are manifested in the topography as relatively elevated hummocks with traces of modern denudation, such as the Inder (a), Zhaman-Inder (a'), Zhundikuduk (b), and other salt dome uplifts (Figure 11.5) and by actively downwarping salt-withdrawal synclines, manifested in the terrain by marked depressions filled with solonchaks or lakes (Inder Lake and Zhaltyr Lake).

Slightly active local morphological structures are difficult to spot on orbital photographs. Essentially, this can be done only by looking at the characteristic pattern of streams and, in part, at photographic hue. In these cases, the ability to interpret morphostructure using remote images of Earth's surface is greatly enhanced when performed in tandem with expert analysis and morphometric study of topographic maps.

Morphostructural analysis of photographs in combination with topographic maps is also helpful in identifying large block-type morphostructures associated with the most recent tectonic activation of basement blocks. In the northern Caspian area, such block morphological structures are denoted primarily by differences in relief

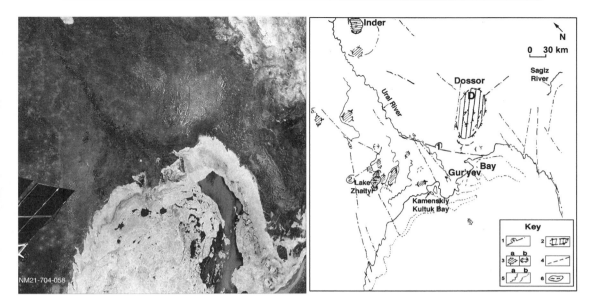

Figure 11.2 Portion of Ural River delta, March 1996 (NASA photograph NM21-704-58), and geomorphological interpretation. Symbols are the same as for Figure 11.1.

within the confines of blocks, which are generally bounded by regional morphological lineaments. The latter are readily interpreted on photographs, but photographs usually do not indicate whether these morphological lineaments bound or intersect the blocks. In the lower Caspian coast, block morphological structures are not elevated relative to each other by very much (≤ 8 m) and are hardly visible in the local topography or in photographs. The elevation changes can, however, be identified through morphometric analysis of topographic maps.

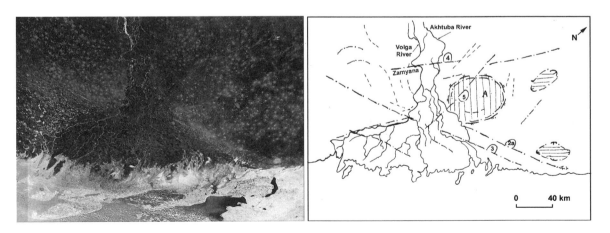

Figure 11.3 Lower Volga River delta, March 1996 (NASA photograph NM21-704-56), and photointerpretation. Symbols are the same as for Figure 11.1.

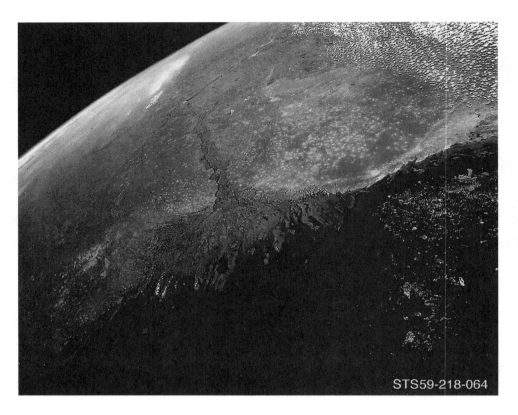

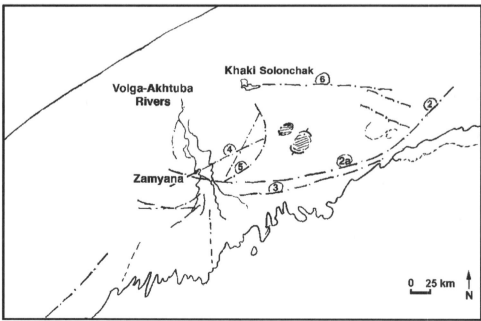

Figure 11.4 Area between the Volga River delta and the top of Ural River delta, April 14, 1994 (NASA photograph STS059-218-64, see also Figure 19 color insert), and photointerpretation. Symbols are the same as for Figure 11.1.

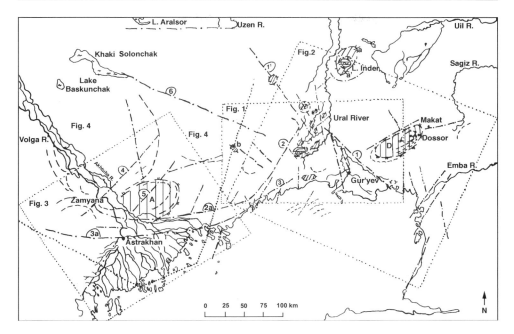

Figure 11.5 Morphostructural diagram from a topographic map based on photointerpretations of space photographs in Figures 11.1 to 11.4. These interpretations are combined to show the structures for the entire segment of the northern coast of the Caspian Sea. Symbols are the same as for Figure 11.1.

11.2.2 Sedimentary Cover

The northern coast of the Caspian Sea is a region covered by a thick, virtually undisturbed stratum of ancient Caspian deposits taking the form of a vast blanket that can only be studied from drilling samples. Only on individual segments in the valleys of major rivers, primarily the Volga and Ural, can a detailed structural study be performed on natural cross sections of blanket deposits, the bulk of which are comprised of sediments from ancient transgressions of the Caspian. One is the segment of Volga valley between the Seroglazovka and Selitryanoe settlements (Figure 11.6 see the color insert).

Photograph STS064-101-20 (Figure 11.6) shows the geological structure of the upper part of the most recent sedimentary cover in the Volga valley. The surrounding plain is covered by a thin (2 to 3 m) blanket of upper Khvalyn (Pleistocene) deposits composed of sandy and grayish-brown dusty sand with isolated pockets of Khvalyn mollusks. The upper part of the blanket is usually subject to aeolian processes and exhibits an irregular surface consisting of semistabilized to unstabilized sands and depressions. Upper Khvalyn deposits (1) can be readily interpreted from photographs on the basis of the gray color with numerous light patches locally corresponding to hills of windblown sand.

The Volga valley cutting through the upper Khvalyn plain appears on the photograph as a wide, sublatitudinal band from which the main elements of its geological and geophysical structure can readily be interpreted (Figure 11.6): streambeds of the primary Volga channels filled with fluvial deposits (2); modern beach deposits, bars and spits, composed of light gray sand (3); and the lower floodplain composed of stratified silty sands partially overgrown with grassy vegetation, and including numerous small tributaries (4).

Significant portions of the valley are in the upper floodplain (5), the primary agricultural production site in the Volga River valley. Its geological structure is like that of the lower floodplain except for considerably reworked floodplain soils encountered repeatedly in the upper floodplain. Areas of ancient deposits in the valley, generally thin sandy and silty deposits with biogenic residues and oxbow lakes, can also be identified (6).

The structure of more ancient strata underlying sediments from post-Khvalyn transgressions and Volga fluvial deposition is revealed in the numerous outcrops of the Seroglazovka and Selitryanoe settlements, visible on the photograph as segments of the post-Khvalyn plain cut through by the Volga and Akhtuba rivers.

11.3 CONCLUSION

Space Shuttle photographs were used to obtain additional morphostructural information about the deep-seated structure and sedimentary cover of the northern coast of the Caspian Sea. The primary result of interpretation of orbital photographs obtained under the joint Shuttle–*Mir* program was the detection of long-lived ruptures and fracture zones that are expressed as regional morphological lineaments, and of two large morphological structures indicative of uplift of subsalt strata.

Photograph STS064-101-20 depicts the structure of young deposits that constitute the upper part of the Pliocene–Quaternary cover. Of these, the alluvial formations of the Volga valley and the post-Khvalyn (post-Pleistocene) marine transgressive deposits are mapped with greatest precision. The photograph also reveals sites with the most eroded sections of the Khvalyn plain, where reference sections of ancient Caspian deposits are being discovered beneath the most recent sediments.

ACKNOWLEDGMENTS

The authors acknowledge the efforts of TechTrans International, Inc. (Houston) in translating this manuscript into English, L. Prejean and C. A. Evans for assistance with illustrations, and M. R. Helfert and staff members of the JSC Office of Earth Sciences for constructive reviews of the paper.

REFERENCES

Aristarkhova, L. B. 1981. Role of geomorphological criteria in revealing deep-seated fractures in "hidden" platform regions (based on the Caspian depression) [original in Russian]. *Geomorfologiya (Geomorphology)*, 1:41–50.

Aristarkhova, L. B., and Fedorova, N. A. 1982. New data concerning the most recent morphological structure and deep-seated structure of the Ryn-Peski Massif (in conjunction with oil and gas exploration) [original in Russian]. *Geomorfologiya (Geomorphology)*, 4:43–51.

Aristarkhova, L. B., and Pryakhina, E. A. 1994. Deep-seated structure of Caspian depression from the standpoint of geomorphological and geophysical data [original in Russian]. *Geomorfologiya (Geomorphology)*, 4:44–52.

Aristarkhova, L. B., Berzin, R. G., and Kerimova, I. K. 1991. Reconstruction of topography and morpholithogenesis of the late Paleozoic in the Caspian via the seismostratigraphic analysis of regional seismic profiles [original in Russian]. *Geomorfologiya (Geomorphology)*, 4:57–63.

Slepakova, G. I. 1977. Continuation of the Pachelm aulacogen in the Caspian depression [original in Russian]. *Geotektonika (Geotectonics)*, 3:46–50.

Zhuravlev, V. S., and Kuzmin, Yu. Ya. 1960. The proposed Azgir fracture in the southern part of the Ural and Volga interfluve [original in Russian]. *Doklady Akademii Nauk CCCP (Reports from the Soviet Academy of Sciences)*, 130(2).

Chapter 12

Shoreline Dynamics and the Hydrographic System of the Volga Delta

N. I. Alekseevskiy, D. N. Aibulatov, and S. V. Chistov

Department of Geography
Moscow State University
Moscow, Russia

ABSTRACT

Coastline change in the vicinity of the Volga River delta has been investigated through use of satellite images, maps, and published hydrological data. In the past 200 years, the delta has built out unevenly, with greater development along the delta's eastern channels throughout the twentieth century. From the end of the nineteenth century to the present, the dominant process has been concentration of flow in progressively fewer major channels and loss of smaller distributaries. Historical coastlines are examined for insight into possible coastal changes if the Caspian sea levels continue to rise.

12.1 INTRODUCTION

The Volga delta is characterized by multiyear changes in shoreline position and in the structure of the hydrographic system. These processes have been accompanied by redistribution of water and sediment among distributaries. Recent growth in the delta area has been dominated by flow concentration in a limited number of the largest main delta channels and the disappearance of numerous smaller distributaries. Factors leading to redistribution of flow into large channels and the abandonment of distributaries include (1) the uneven advance of the delta toward the sea in various directions, (2) the lowering of Caspian sea level throughout most of the twentieth century (during periods of sharp decline in sea level the concentration of flow into larger channels near the sea increased), and (3) the artificial deepening of the estuaries of some streams.

Volga delta evolution has been fairly well studied. Nevertheless, the appearance of new data and the development of new analytical methods for cartographic data permit expanded research on the dynamics of the deltaic hydrographic system and its shoreline. It is important to understand historical changes in the delta as analogs for predicting the changing position of delta and shoreline under future sea-level changes. One way to identify situations in the past is to overlay maps and satellite imagery from various time periods.

12.2 METHODS

To evaluate the dynamics of the Volga delta coastline in the nineteenth century, we used data on shoreline shifts from various authors (State Admiralty Department, 1826; Baidin, 1962; Mikhailov et al., 1986). These evaluations are rough approximations. However, they permit us to evaluate in a general way the change of the delta shoreline and the morphology of islands in the coastal estuary.

More information is available on changes in the Volga delta during the twentieth century. We used Russian cartographic materials from 1914, 1920, 1935, and 1941, Landsat imagery from 1978, Soviet satellite imagery from 1991, and NASA astronaut photographs from 1996. We performed image analysis and transformation of the contours using the Pericolor-3000 system. All maps and space imagery were co-registered to one form and scale of presentation, based on the selection of a system of tie points on the maps and pictures for different years. This co-registration allowed us to map the change in position of the coastline for each map and photograph separately, and then on the overall map.

We estimated the degree of change to the delta system channels using a metric, D, which is equal to the ratio of the total width of all the delta channels in a region to the area of that region (Alekseevskiy et al., 1997). This value is a function of the maturity of lobe development in different regions of the delta—the older a delta lobe, the smaller is D.

The value of the D is also seasonal, depending on the volume of discharge at the head of the delta. During periods of flooding D is greater (wider channels), and in periods of low flow, D is smaller. In this chapter, changes in D are considered over long time intervals and during stages of lowest flow, when the influence of discharge on the D is minimal. Changes in D indicate that the structure of the hydrographic network of streams over the whole delta has changed. As the delta shoreline advances, upstream regions experience a decrease in D. In contrast, close to the shoreline the number of distributaries increases, as do values of D.

12.3 RESULTS

The position of the Volga delta shoreline is determined by many factors, one of which is sea level. When sea level drops, the shoreline generally moves passively toward the sea; when sea level rises, the shoreline moves toward dry land. Another cause for the seaward advance of the shoreline is the accumulation of river sediment. Analysis of combined historical maps, the fluctuations of Caspian sea level (Figure 12.1), and data on flow rates and sedimentation in the delta (Table 12.1) allow us to

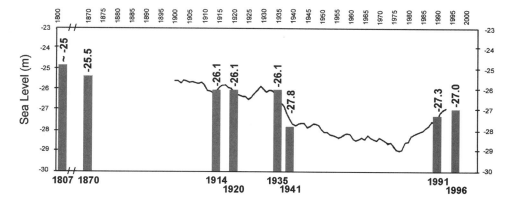

Figure 12.1 Benchmarks of Caspian sea level linked to cartographic and satellite materials used in this study. Bars indicate sea level in the Volga delta region reported by the authors. Line indicates data for the twentieth century used by Varuschenko et al. (Chapter 16).

infer the direction of delta growth for several time intervals. We are also able to examine major factors affecting delta changes, such as dams built on the Volga between 1940 and 1960.

12.3.1 Shoreline Dynamics

The formation of the Volga delta in the last 200 years has taken place against the general backdrop of a lowering sea level (Figure 12.1). Present Caspian sea level is −26.9 m (Chapter 9). Rychagov (1994) believed that the sea level in the last few hundred years has not exceeded −25.0 m. Falling sea level was accompanied by significant passive advance of the delta shoreline.

Figure 12.2 is a map of the Volga delta showing some of the main channels (labeled on a ca. 1991 base), and Figure 12.3 is a compilation of Volga delta coastlines over time. Initially, the western portion of the delta consisted of three debris cones: the Volga itself, the Bakhtemir, and the Staraya (Old) Volga (State Admiralty Department, 1826; Baidin, 1962). Over the last 200 years, the most active delta for-

TABLE 12.1 Volga Delta Formation and Shoreline Dynamics Based on Historical Maps and Space Imagery

Period	Average Annual Flow Rate (m^3/yr)	Average Annual Sedimentation Rate ($\times 10^6$ metric tons/yr)	Sea-Level Change (m)	Rate of Delta Advance (km/yr)			
				East	Center	West	Average
1807–1870	249	7.7	−1.0	0.7	0.5	0.2	0.47
1870–1914	250	8.0	−0.6	—	—	—	0.21
1914–1920	286	8.8	0	—	—	—	0.19
1920–1935	261	8.1	−0.02	0.5	0.0	0.4	0.30
1935–1996	236	7.6	−1.07	0.2	0.2	0.3	0.23

mation occurred along the lobes in the western part of the delta, along the Volga, the Bakhtemir, and the Kamyzyak channel systems. Minimal advances occurred in the eastern and central delta along the Buzan, Shmagina, and B. Bolda branches (Figure 12.3). Very likely, the uneven advance of the delta was related to the development of a system of distributaries in the western portion of the delta and the variable slope of the increasingly emergent seafloor. Alluvium filled the areas between the hills, and the river cut through to the bottom of the bay. A number of islands existed beyond the delta margin, in the areas of the present western and eastern *ilmen* (Volga delta lakes), where river sediments accumulated. The Buzan channel extended between lobes in the east and in the southwest. The flow direction and the formation of large deposits at the site of former *kultuks* (deeply indented shallow bays) indicate that the waters were deeper in the eastern portion than in the western portion of the estuary in the past.

The position of the delta shoreline in the early nineteenth century (Figure 12.3) can be used to infer that if sea level rose again to −25.0 m, there would be a larger transgression of the present shoreline in the eastern portion of the delta. We recognize that the accuracy of a paleohydrographic reconstruction of the dynamics of the delta shoreline at a sea level of −25 m is limited. The coastline position at the beginning of the nineteenth century (a relative highstand) and at the end of the twentieth century differ because of both subaerial and subaqueous delta growth and sediment redistribution in the past 200 years. However, in the western portion of the delta, the width of the delta flood plain is 15 km, and in the eastern delta, it is 45 to 50 km. With some allowances, we estimate that the delta would be located 15 to 20 km north of the −25.0-m structure contour (paleo-waterline).

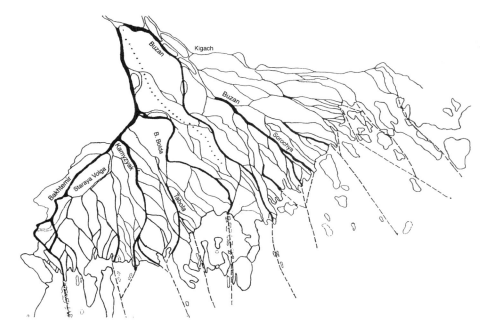

Figure 12.2 Map of the main channels in the Volga delta, labeled on a ca. 1991 map. The dashed lines coming radially from the delta shoreline are artificially maintained offshore channels.

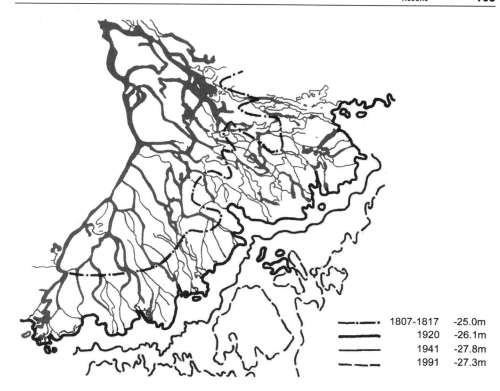

Figure 12.3 Position of the Volga delta marine edge in 1807 to 1817, 1920, 1941, and in 1991 with Caspian sea levels of −25.0, −26.1, −27.8, and −27.3 m, respectively. The shoreline positions are mapped on a base map of distributary channels in 1920.

From 1807 to 1870 (Table 12.1) the decrease in the sea level by a little more than 1 m (against the backdrop of an average annual inflow of 249 m³ of water and 7.7 million metric tons of alluvium into the estuary region) led to a significant but mostly passive advance of the delta. In the eastern portion the advance was 42 km, in the central it was approximately 30 km, and in the western it was 13 km. The average rate of the advance was 0.7, 0.5, and 0.2 km/yr, respectively (Table 12.1).

In 1861 the western portion of the territory had several main arms: Bakhtemir, Staraya Volga, Kamyzyak, and B. Bolda. Each of these had its own system of branching channels. The western delta shoreline was extremely rugged, with a small number of little islands (Figure 12.4). The distributary system in the eastern delta became significantly more complicated at that time. This may be indicative of the activation of the Buzan channel due to the significant elongation of the delta lobe in other flow directions. A large number of islands of varying sizes and shapes were formed beyond the estuary.

For the period 1870 to 1914, sea level decreased to −26.1 m (Figures 12.1 and 12.3). Other hydrographic conditions were similar to those from 1807 to 1870. During that period the distributary systems of Bakhtemir, Staraya Volga, Kamyzyak, and B. Bolda continued to develop. Intense flow segmentation in the coastal area was observed. The Bakhtemir estuary advanced seaward 4.5 km (the largest advance rel-

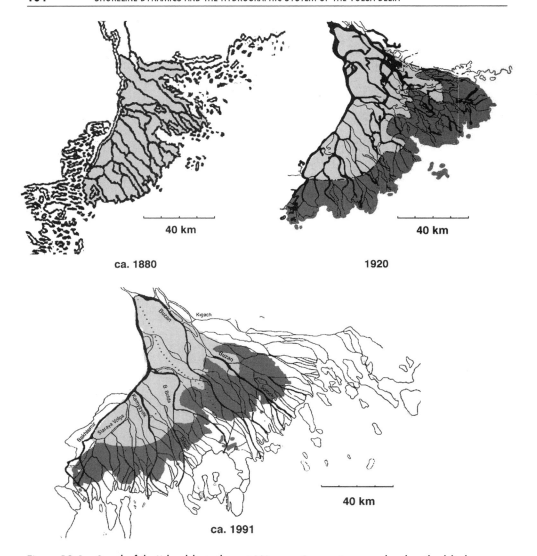

Figure 12.4 Growth of the Volga delta in the past 100 years. Comparative maps show how the delta has grown unevenly over time (from west to east), and the relative position of the shoreline. Base maps shown from left to right: ca. 1880 map (from *Comprehensive Atlas and Geography of the World*, 1882); 1920 map (from Figure 12.3), ca. 1991 map (from Figure 12.2). The nineteenth-century delta (light shading) and the delta growth by 1920 (dark shading) are highlighted and superimposed to facilitate comparisons.

ative to the other channels), indicating that it received a large percentage of Volga water and sediment. The eastern shoreline of the delta began to advance toward the sea rapidly due to the activation of the Buzan and the Shmagina and Kigach arms. The coastline remained very rugged, but beyond the delta shoreline the number of small islands sharply decreased and large islands disappeared completely, attesting to the gradual filling of the coastal estuary with fluvial sediments and to the coalescence of large islands into the delta.

In the period 1914 to 1920, sea level remained stable; the average annual inflow of water and alluvial deposits was 286 m³ and 8.8 million metric tons, respectively. The delta advanced at a rate of 0.19 km/yr, a twofold decrease compared with the preceding period. These estimates closely coincide with Baidin's (1962) data. The structure of the delta hydrographic system continued to become more complicated. The position of the delta shoreline at −26.1 m is shown in Figure 12.3, which reflects the situation in 1920.

In the following years, 1920 to 1935, under conditions of decreasing flow and sediment transport (Table 12.1), the delta area increased 1800 km² (including 700 km² of islands). The total average seaward advance was 9.5 km (Mikhailov et al., 1993). Most of the delta growth occurred along the western and eastern arms. In the central delta the shoreline position remained practically stable (Table 12.2 and Figures 12.3 and 12.4).

After 1935 the advance of the delta slowed sharply; the position of the delta shoreline stabilized. The total increase in area was only 50 km². This is explained (Mikhailov et al., 1993) by the fact that a large portion of the present shallow zone of the coastal estuary is located above the −29-m mark, and therefore the delta shoreline contours at levels lower than the −29.5 to −28 m are not determined by the sea-level position but by other factors, such as the Volga floodwaters. A specific "buffer" zone arose between the delta and the sea. The natural lowering of water volume in the Volga River and the anthropogenic decrease in suspended sediment load significantly affected this process.

From 1936 to 1996, sea level fell 1.07 m. The period was distinguished by minimal river flow (the average annual water flow was equal to 236 km³; suspended sediment load was 7.6 million metric tons; Table 12.1). Due to these factors, the rate of advance varied from 0.2 (eastern and central portions) to 0.3 km/yr (western portion). These estimates agree very well with the data presented by Mikhailov and co-authors (1993). Although some coastal lowlands have flooded recently, the sea-level rise from 1978 to 1996 to between −27.5 and −27.0 m did not result in a significant landward shift of the delta, due to the morphology of the coastal estuary (Figure 12.5, see also Figure 19.2 in color insert).

Table 12.2 Shoreline Advance in the Estuary Region of Volga Delta Channels, Based on Cartographic Materials and Satellite Imagery[a]

	Estuary Elongation (km)	
Channel	1920–1935	1936–1978
Bakhtemir	6.9	17.2
Koklyui	0.5	14.7
Biryul	0.0	8.5
Ivanchug	0.0	18.2
Kamyzyak	1.6	14.2
B. Bolda	0.3	11.2
Sorochya	0.0	14.5
Buzan	1.6	8.8
Kigach	15.0	20.9

[a] Since 1979 the shoreline has retreated due to transgression of the Caspian Sea.

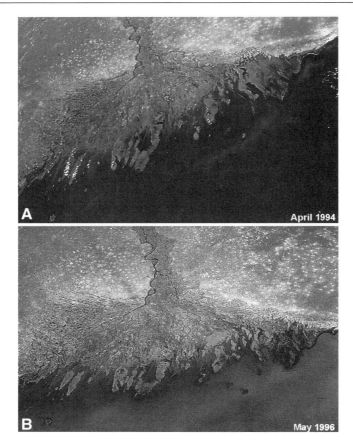

Figure 12.5 The Volga Delta in (*A*) April 1994 (detail of NASA photograph STS059-218-64, see also Figure 19.2 in the color insert) and (*B*) May 1996 (NASA photograph NM21-735-42), showing continued transgression. Slight differences in coastline position can be identified. Compare the May 1996 shoreline with the map in Figure 12.2.

12.3.2 Evolution of Volga Delta Channels

The general evolution of the delta is closely connected with the processes of reconstructing the channel system. The disappearance of distributaries in the Volga delta has been confirmed through an analysis of changes in the total number of channels (Baidin, 1962; Mikhailov et al., 1986; Polonskiy et al., 1992). Despite the progressive increase in the number of distributaries in the lowest portion of the delta (from 200 at the end of the nineteenth century to 475 in 1930, 800 in 1960, and 1000 by 1980; Mikhailov et al., 1977; Polonskiy et al., 1992), the number of channels in the Volga delta decreased during the twentieth century as abandonment outpaced the creation of new delta channels.

From 1937 to 1939 there were a total of 1107 channels in the Volga delta (Baidin et al., 1956): 71 in the upper delta, 223 in the middle, and 813 in the lower portion of the delta. The total length of these channels in the delta system exceeded 7500 km (Baidin, 1962). By the mid-1970s the number of channels that transited the delta decreased to 757 (small ones in channel-mouth bars in estuaries at the delta shoreline are not counted), with a total length of 4500 km (Mikhailov et al., 1977). Recognizing that these estimates are approximate and that there are difficulties in comparing counts from different periods, we still conclude that there has been widescale abandonment of channels in the Volga delta.

The results of an analysis of the variability of D (where D equals the ratio of channel width to lobe area) showed that during the period 1914 to 1935, a zone of active distributaries became substantially wider and moved seaward; for most regions in the delta, $D > 30$ m/km². Almost all of the eastern delta, including the Buzan interdistributary system and the eastern channels of the B. Bolda system, became increasingly complex. A similar situation existed in the zone where the Kizana (Kamyzyak) and Staraya (Old) Volga channel systems formed. The coastal sector of the entire delta was distinguished by increased values for the D, generally greater than 30 m/km². Exceptions were the territories in the lower reaches of the western arms of the B. Bolda and Bakhtemir system, where the D ranged between 10 to 20 m/km². $D > 30$ m/km² was characteristic for portions of the delta adjacent to Bakhtemir for roughly half its length.

After 1935, development of the delta hydrographic system reflected increased flow regulation upstream. A decrease in the water flow during the flood period led to a general lowering of the value of D, not exceeding 10 to 20 m/km². The areas with these D values included the Staraya (Old) Volga, Kizana (Kamyzyak), and B. Bolda channel systems. During this time the Buzan channel developed a relatively large concentration of delta distributaries, indirectly reflecting the direction of sediment accumulation.

From 1935 to 1991, the degree of channel dissection sharply declined. Only some portions of the coastal sector—the central and upper delta—had D values greater than in 1935. Although total river flow increased in 1991, there was average or lower-than-average flow during most of this time, and the D value (relative to 1935) was even lower.

12.4 CONCLUSION

Using maps, satellite images, and space photographs, we have examined coastline changes in the vicinity of the Volga delta. This work was based on detailed information captured in the images, on knowledge of the delta accretionary processes, and on historical reconstructions during highstands of Caspian sea level. Over the past 200 years, the delta has advanced asymmetrically (Figure 12.4), due to both sea-level drops and variable distributary development. Much of the recent growth has occurred in the eastern distributary systems. An analysis of channel development indicates that flow has, in general, been progressively concentrated in the larger channels at the expense of many smaller distributaries.

Results can be used to predict future shifts in the delta shoreline and the hydrographic system at even higher sea levels based on historical analogs. Predicting the probable future position of the delta shoreline at a sea level around −25.0 or −26.0 m is important. In the event of a sea-level rise to −25.0 m, many cultural and industrial facilities may be destroyed or flooded. Engineering projects are necessary to protect these facilities and the 15 Volga delta villages with a total population of approximately 40,000 people.

We identified historical analogs in the situation at the beginning of the nineteenth century, when sea level was around −25.0 m, and at the beginning of the twentieth century, when it was stabilized at about −26.0 m (Figure 12.3). Complete reconstruction of the shoreline position is not possible, as the elevation of the delta surface and its morphology have changed with time. However, the general position of the delta coastline would correspond roughly to the situation depicted in Figure 12.3. Thus, it is possible to plan procedures for reducing the flood hazards the Volga delta.

ACKNOWLEDGMENTS

The authors acknowledge the efforts of TechTrans International, Inc. in translating this manuscript into English. Reviews and figure preparation by staff members of the JSC Office of Earth Sciences are also acknowledged.

REFERENCES

Alekseevskiy, N. I., Aibulatov, D. N., Chistov, S. V., Sventek, U. V., and Serapinas, B. B. 1997. The dynamics of the Volga delta marine margin. In Riverbed Processes in the Volga Delta. Geoecology of the Caspian Region Series [original in Russian]. Moscow, pp. 140–155.

Baidin, S. S. 1962. *Stok i urovni del'ti Volgi (Flow and Levels of Volga Delta)*. Moscow: Gidrometeorologicheskoe izd-vo, 337 pp.

Baidin, S. S., Linberg, F. N., and Samoilov, I. V. 1956. *Volga delta hydrology* [original in Russian]. Moscow: Gidrometeoizdat, 331 pp.

Comprehensive Atlas and Geography of the World. 1882. Edinburgh, Scotland: Blackie & Sons, Available at Federation of East European Family History Societies map room index at *http://feefhs.org/maps/ruse/re-riv.html* (viewed January 14, 1999)

Mikhailov, V. N., Rogov, M. M., Makarova, T. A., and Polonskiy, V. F. 1977. *Dinamika gidrograficheskoæi seti neprilivnykh ust'ev rek (Dynamics of the Hydrographic System of Nontidal River Estuaries)*. Moscow: Gidrometeoizdat, 294 pp.

Mikhailov, V. N., Rogov, M. M., and Chistyakov, A. A. 1986. *Rechnye del'ti: gidrologo-morfologicheskie protsessy (River Deltas: Hydromorphological Processes)*. Leningrad: Gidrometeoizdat, 280 pp.

Mikhailov, V. N., Korotaev, V. N., et al. 1993. Hydro-morphological processes in the Volga estuary region and their changes under the influence of Caspian sea level fluctuations [original in Russian]. *Geomorfologiya (Geomorphology)* 4:97–107.

Polonskiy, V. F., Lupachev, Yu. V., and Skriptunov, N. A. 1992. *Gidrologo-morfologicheskie protsessy v ust'yakh rek i metody ikh rascheta (prognoza) [Hydro-morphological Processes in River Estuaries and Methods for Calculating (Predicting) Them]*. Saint Petersburg: Gidrometeoizdat, 383 pp.

Rychagov, G. I. 1994. Caspian sea level at the end of the 18th century and beginning of the 19th century [original in Russian]. *Geomorfologiya (Geomorphology)* 1:102–108.

State Admiralty Department. 1826. Drawings using the descriptions and astronomical observations made during 1809–1817 by 8th class Navigator and Cavalier Kolodkin. In *Atlas of the Caspian Sea*.

Chapter

13

Changes in Avian Habitats in Volga Delta Wetlands During Caspian Sea-Level Fluctuations

E. A. Baldina and I. A. Labutina
Geography Department
Moscow State University
Moscow, Russia

G. M. Rusanov, A. K. Gorbunov, and A. F. Zhivoglyad
Astrakhanskiy Biosphere Reserve
Astrakhan, Russia

J. de Leeuw
International Institute for Aerospace Survey and Earth Sciences
Enschede, The Netherlands

ABSTRACT

The wetlands of the Volga River delta have been designated wetlands of international importance and include the protected areas of the Astrakhanskiy Biosphere Reserve (*Zapovednik*). Space photographs help document the effects of sea-level change in the Caspian Sea on the vegetation of the delta and on habitat suitability for waterbirds and mammals. The dramatic increase in vegetation that accompanied the most recent regression (until 1978) attracted great flocks of birds that nested and/or wintered in the area. Sea-level rise (1978 to present) and increased Volga River runoff have changed the quantity and distribution of aquatic vegetation. During the present highstand, extensive mats of floating debris serve as habitat for birds and some mammals, but overall habitat availability and quality has decreased.

13.1 INTRODUCTION

The Volga delta is one of the largest deltas in the world. Its land surface occupies an area of approximately 10,000 km^2. This is joined by the underwater portion, which forms a shallow-water prodelta with a total area of approximately 5500 km² (Kroonenberg et al., 1997). Under the 1971 Ramsar Convention on wetlands conservation, the Volga delta was included in the list of wetlands having international importance as a waterfowl habitat (Carp, 1972; Smart, 1976).

The Volga delta wetlands are key nodes in the migration paths for many species of birds from western Asia, North Africa, and the Mediterranean to western Siberia and the northern European area of Russia. In addition, they serve as a breeding and molting site for a large avian population (Isakov and Krivonosov, 1969).

The primary lands requiring protection occupy the lower subaerial delta and the shallow offshore prodelta. A photograph from space shows the extent of this region, covered with emergent and submerged vegetation, with numerous shallow islands (Figure 13.1, Figure 19.2 in color inserts). It extends laterally for approximately 200 km and is 40 to 60 km long.

The delta wetlands are located in the zone of influence of both the Caspian Sea and the Volga River. Caspian sea level fluctuates significantly, and the Volga River drainage system is also subject to significant variations (Figure 13.2). This has a significant impact on the delta wetlands.

13.2 METHODS

We analyzed several space photographs and satellite images and summarized analyses using a geographic information system (GIS). Remotely sensed data are impor-

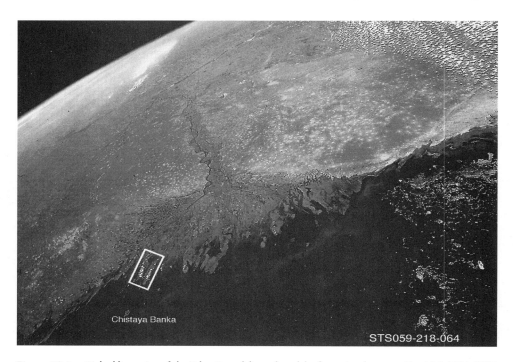

Figure 13.1 High oblique view of the Volga River delta and prodelta for regional context, (April 14, 1994, NASA photograph STS059-218-64, see also Figure 19.2 in the color insert). The boxed area is the Damchik section of the Astrakhanskiy Biosphere Reserve shown in Figures 13.4 and 13.5.

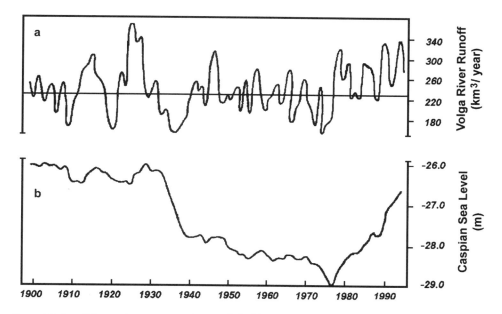

Figure 13.2 Multiyear changes in (*a*) Volga runoff (km³/yr), and (*b*) Caspian sea level (m). (Modified from Baidin and Kosarev, 1986.)

tant mapping and monitoring tools for the rapidly changing conditions of wetlands regions. The effectiveness of their use is improved when advanced geoinformation technologies are used (Soeters et al., 1991; Saraf and Choudhury, 1998). These methods are exceptionally valuable in studying and monitoring river delta ecosystems, which are distinguished by their high degree of variability.

Three sections of the Volga delta were studied in detail. The Astrakhanskiy Biosphere Reserve GIS was created for comprehensive studies of the reserve and for administration of the regions of international cooperation. The reserve is designated United Nations Category IV and comprises 63,400 ha (United Nations Environmental Programme, 1993). Several regions within the reserve were studied (Figure 13.3); these sections encompass the most typical ecosystems of the delta and can be viewed as a test area for the entire Volga delta (Lychagin et al., 1995). We provide results from the Damchik Section and Chistaya Banka (Figure 13.3).

Aerospace data, including both Russian and U.S. space photographs from various years, form the basis for the GIS (Baldina et al., 1995). Our data from space surveys cover the period beginning in the mid-1970s, including the Caspian lowstand and the subsequent Caspian transgression. Photographs obtained in the early and middle 1990s characterize the period of highest Caspian sea level. The primary informational layers of the GIS were created using these data and field data.

One of the primary layers of the GIS is the vegetation map, which reflects the state of the vegetation in the early 1990s (Labutina et al., 1995). Vegetation plays a major role in the functioning of the ecosystems of the reserve and serves as a good indicator of spatial structure and habitat conditions. In addition, the vegetation in this territory is the best indicator of physical change in the environment and is clearly reflected in the photographs. The avian population and its distribution are closely tied to the state of

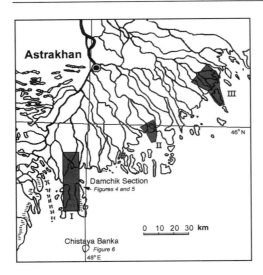

Figure 13.3 Location of sections of the Astrakhanskiy Biosphere Reserve. Areas discussed in Figures 13.4 to 13.6 are marked.

the vegetation as well as to water depths (Rusanov, 1993). The availability of remote-sensing data comprising a time series makes it possible to create maps that characterize the state of the vegetation over time. Such maps serve as a basis for a system for monitoring the Volga delta wetlands. Figure 13.4 (see the color insert) is an example of a vegetation map series of the Damchik section of the Astrakhanskiy Biosphere Reserve.

In this chapter we focus on the changes in vegetation associated with Caspian sea-level change. We also refer to general results from accompanying data that we collected on bird and mammal abundance during field surveys. The details of field methods and survey results are not presented here. Summary data on the flora and fauna incidence in the Astrakhanskiy Biosphere Reserve have been made available through the Man and the Biosphere Program (Information Center for the Environment, 1997, 1998).

13.3 RESULTS AND DISCUSSION

13.3.1 Period of Low Caspian Sea Level

After the sharp drop in sea level in the 1930s, the coastal delta—the prodelta—began to be quickly overgrown with aquatic vegetation. By the 1970s, a significant portion of the water area was covered by thickets of common reeds (*Phragmites australis*), small reedmace (*Typha angustifolia*), sacred water lotus (*Nelumbo nucifera*), bur-reed (*Sparganium erectum*), flowering rush (*Butomus umbellatus*), and other emergent vegetation. Plants with floating leaves, such as water fringe (*Nymphoides peltata*) and water chestnut (*Trapa natans*), and submerged vegetation, such as pondweeds (*Potamogeton* spp.), tape grass (*Vallisneria spiralis*), hornwort (*Ceratophyllum demersum*), stonewort (Charophyta [*editor's note*: order Charales]) and green algae (Euchlorophyceae [*editor's note*: probably synonym of Chlorophyceae]) are very widespread.

By 1977, the Caspian sea level had dropped to −29 m (relative to global sea level). Shallows and bars around islands, where the predominant depths at low water were

20 to 40 cm, were overgrown. Even stream channels with depths of 50 to 60 cm were overgrown. Extensive common reed and small reedmace thickets formed on the prodelta islands. Mesophyllic and even xerophyllic grasses, as well as small forests of white willow (*Salix alba*), formed on certain areas of the islands. Harvesting of common reed and hay, as well as other farming, took place on some islands from the late 1960s to the mid-1980s.

The islands were populated with wild boar (*Sus scrofa*), raccoon dog (*Nyctereutes procyonoides*), European water vole (*Arvicola terrestris*), Norway rat (*Rattus norvegicus*), and other mammal species.

The extensive propagation and varied floral composition of aquatic vegetation in the prodelta under low-water conditions created favorable feeding conditions for many avian species, primarily dabbling and diving ducks, mute swans (*Cygnus olor*) and whooper swans (*C. cygnus*), greylag geese (*Anser anser*), and common coots (*Fulica atra*). The lowstand was characterized by the most favorable nesting and protective conditions for many species of waterfowl and other birds, which was accompanied by their rapid population of the wetlands and an increase in their numbers. The wetlands were used intensively by waterfowl throughout the entire ice-free period, and they served as a wintering area under favorable weather and ice conditions.

Habitats used by avian species varied in relation to their life histories. All the vast flocks of greylag geese and whooper swans formed in places where lotus grew, as it is one of the primary foods for these birds in the Volga delta. Water areas with isolated and semimassed thickets of common reed, small reedmace, bur-reed, lotus, and floating and submerged vegetation served as excellent areas for nesting of waterfowl, wading birds (Ciconiiformes), and sparrows (Passeriformes). Dalmatian pelicans (*Pelecanus crispus*) nested in areas of this type. The mass of developing floating vegetation created favorable conditions for nesting of whiskered terns (*Chlidonias hybridus*), white-winged black terns (*Chlidonias leucopterus*), and black terns (*Chlidonias niger*). Massive summer molting of ducks, common coots, greylag geese, mute swans, and other bird species proceeded in these areas. Conditions were also favorable for certain types of mammals, especially muskrat (*Ondatra zibethicus*).

13.3.2 Onset of Caspian Sea-Level Rise

A rise in Caspian sea level began in 1978. Changes in the vegetative cover in the prodelta during this period were ambiguous. On one hand, the continuing formation of new stands of common reed and small reedmace, their thickening, and even the formation of dense stands of common reed in certain areas are distinctly visible in space photographs. The areas of communities including lotus also increased. On the other hand, the increase in depths and the change in wave conditions caused rapid thinning of plants with fragile stems and shallow root systems, such as milfoil (*Myriophyllum* sp.), tape grass, hornwort, stonewort, and bur-reed.

The rise in sea level and the increase in the Volga runoff caused changes in the seasonal overgrowth of aquatic vegetation. Comparative GIS analysis of two sets of photographs from different seasons (during the period when the Caspian Sea was at minimum level and when its level had risen by 2 m) makes it possible to illustrate the changes that occurred. Figure 13.5A and D, shows the limits of propagation of vegetation in the early summer. At that time it consisted primarily of stands of common

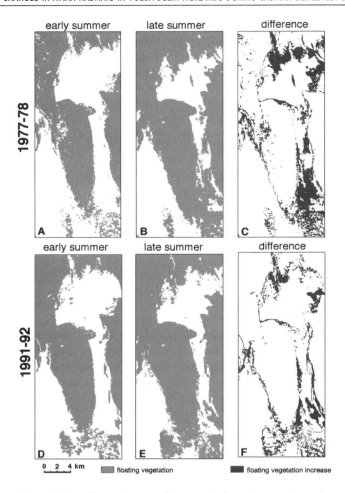

Figure 13.5 Changes in seasonal overgrowth of aquatic vegetation in the Damchik section of Astrakhanskiy Biosphere Reserve: (A) data from classification of Landsat MSS image, June 8, 1977; (B) data from classification of Landsat MSS image, August 29, 1978; (C) difference between vegetation cover in views (A) and (B); (D). data from classification of MSU-E *Resurs O* photograph, June 30, 1991; (E) data from classification of MSU-E *Resurs O* photograph, September 5, 1992; (F) difference between vegetation cover in views (D) and (E). The difference image primarily characterizes the spread of vegetation with floating leaves and thickets of lotus and bur-reed.

reed and bur-reed. Floating and submerged vegetation gradually appeared on the water surface later, and its development was at maximum level in late July and August (Figure 13.5B and E). It is possible to see (Figure 13.5C and D) the magnitude of decrease in area occupied by floating vegetation by the early 1990s; the decrease in floating vegetation has been 33%. As ground observations show, this is explained by the decrease, or nearly complete disappearance, of such aquatic vegetation species as water chestnut, water fringe, and bur-reed due to increased water depths in the prodelta.

The influence of sea-level rise is most distinct in the open prodelta zone. We observed severe disruption of vegetative cover there, especially on the islands. Figure 13.6 shows the results of comparing images of Chistaya Banka Island, which were taken at different times. In the period of low Caspian sea level (1977–1978, Figure 13.6A and B) dense strands of reedmace were observed even in early summer. Red color in the late summer image corresponds with herb and grass meadows that were used as pasture. By 1991 the reedmace thickets around the periphery of the island were destroyed. The meadows in the central portion of the island also disappeared as a result of sea-level rise and were replaced by sparse stands of common reed and Laxmann's reedmace (*Typha laxmannii*).

The change in hydrological conditions also had a negative effect on bird habitat. By the mid-1980s, when sea level had risen by 1 to 1.2 m (reaching absolute levels of –28 to –27.8 m) the most significant changes had occurred in the open prodelta zone, in the southernmost portion of the water area. Molting dabbling ducks reacted first to the changes in the ecological conditions in these areas. Their numbers began to decrease rapidly due to the increase in depths and the deterioration of the feeding and protective properties of these areas. By the early 1990s, these water areas were no longer suitable for mass summer molting of ducks as had occurred previously. Only mute swans and a small number of common coots and greylag geese continued to use them for molting.

The changes in the wetlands during the initial period of sea-level rise had less effect on waterfowl reproduction. The nesting waterfowl population remained high until the mid-1980s. By the late 1980s and early 1990s, the depths offshore had begun to exceed 1 m and reach 1.5 m or more under high-water conditions. Due to the seaward slope of the delta plain, the sea-level rise had differing effects on depth in different parts of the prodelta. The range of water-level fluctuations increased significantly due to wind-induced surges. Changes in ice conditions had an increasingly destructive effect on vegetation. Our observations indicated that all these factors

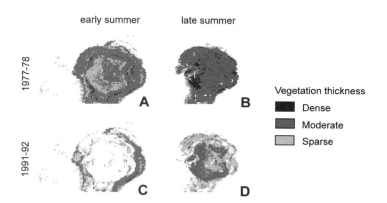

Figure 13.6 Change in vegetation on Chistaya Banka Island during the 1978–1992 period using data from (*A*) Landsat MSS image, June 8, 1977; (*B*) Landsat MSS image, August 29, 1978; (*C*) MSU-E *Resurs 0* photograph, June 30, 1991; and (*D*) MSU-E *Resurs 0* photograph, September 5, 1992. Three vegetation classes are represented based on the Normalized Difference Vegetation Index (NDVI).

degraded bird nesting conditions and were accompanied by a decrease in the breeding bird population. The most ecologically adaptable mute swans and common coots began to populate the new common reed and small reedmace stands on flooded islands, which previously had been unsuitable for their nesting.

The floating vegetation became sparse, depriving terns of nesting places. Nearly all nesting colonies of terns were now confined to wetlands located along the sea edge of the former delta, where the floating vegetation—yellow water lily (*Nuphar lutea*), white water lily (*Nymphaea candida*), water chestnut, and water fringe—densely covered significant areas, creating favorable conditions for terns to breed after the high water dropped in July and August. The numbers of wading birds in the wetlands also decreased sharply.

The habitat for mammals deteriorated. The muskrat population dropped sharply, and many areas become completely unsuitable for other mammal species. Muskrats began to populate the closed reed and reedmace stands on large islands in the prodelta, which were submerged throughout the year.

13.3.3 Caspian Sea-Level Highstand

The most recent space photographs (Figure 13.7, below, and Figure 19.3, color insert) make it possible to characterize the present state of the prodelta wetlands. The very low islands are flooded, and vegetation in the southern portion of the region has become very sparse.

Figure 13.7 State of Volga delta wetlands during period of high Caspian Sea level, September 1994 (NASA photograph STS064-101-17, see also Figure 19.3 in the color insert). The boxed area is the Damchik section is discussed in Figures 13.4 and 13.5.

By the mid-1990s water depths there reached 1.6 to 1.8 m at high water, and the wetlands finally lost their importance as a site for mass breeding of waterfowl. Isolated and aggregated stands of reed and reedmace, which previously had been the best nesting areas, were destroyed by changing water and ice conditions and became completely unsuitable for nesting by many avian species.

During the winter, vast quantities of reeds are torn loose by ice and carried out to sea. The plant material is then forced by wind and waves into the peripheral sections of the closed reed and reedmace stands on former islands, forming thick mats of floating plant debris that may extend for 20 km or more. Our ground and aerial surveys of the wetlands have shown that birds continue to nest only in these floating mats of plant debris. Mute swans and Dalmatian pelicans build their nests here. Debris islands serve as a habitat not only for birds but also for mammals—muskrat and, more rarely, raccoon dog and American mink (*Mustela vison*). Our observations of migrating birds show that further changes are occurring in their populations. Open-water birds (swans and diving ducks) are increasing in percentage, while the number of dabbling ducks and greylag geese is decreasing.

CONCLUSION

The regular acquisition of new remote-sensing data together with ground and aerial visual observations may serve as the basis for ongoing monitoring of wetlands in the Volga delta. The link of sea-level rise to vegetation change and bird abundance will provide important information that can be used to better manage protected areas.

ACKNOWLEDGMENTS

The authors acknowledge TechTrans International, Inc. (Houston) for translating the manuscript into English. Helpful reviews by M. R. Helfert and staff of the JSC Office of Earth Sciences are also acknowledged. The assistance of J. A. Robinson and L. Prejean with illustrations is appreciated.

REFERENCES

Baidin, S. S., and Kosarev, A. H., eds. 1986. *Kaspiiskoe more: gidrologiya i gidrokhimiya (Caspian Sea: Hydrology and Hydrochemistry)*. Moscow: Nauka, 261 pp.

Baldina, E. A., Knizhnikov, Y. F., and Labutina, I. A. 1995. The Astrakhanskiy Biosphere Reserve GIS. Part 2: Aerospace and cartographic maintenance. *ITC Journal*, 3:193–196.

Carp, E., ed. 1972. *Proceedings, International Conference on the Conservation of Wetlands and Waterfowl*, Ramsar, Iran, January 30–February 3, 1971. Slimbridge, Gloucester, England: International Waterfowl Research Bureau, 303 pp.

Information Center for the Environment and Man and the Biosphere Program. 1997. MAB Biosphere Reserves in the Russian Federation: Astrakhanskiy, MAB-Fauna Database, species lists for Astrakhanskiy Biosphere Reserve *http://endeavor.des.ucdavis.edu/mab/reserve.asp?reserve=RUSAST00* (viewed February 5, 1999)

Information Center for the Environment and Man and the Biosphere Program. 1998. MAB Biosphere Reserves in the Russian Federation: Astrakhanskiy, plant species reported from the world's protected areas, 313 plant species in Astrakhanskiy Biosphere Reserve. *http://endeavor.des.ucdavis.edu/mab/flora.asp?reserve=RUSAST00* (viewed February 5, 1999)

Isakov, Yu. A., and Krivonosov, G. A. 1969. Waterfowl migration and molt in the Volga delta, Astrakhan [original in Russian]. *Proceedings, Astrakhanskiy State Nature Reserve,* Vol. 12, 186 pp.

Kroonenberg, S. B., Rusakov, G. V., and Svitoch, A. A. 1997. The wandering of the Volga delta: A response to rapid Caspian sea-level change. *Sedimentary Geology,* 107:189–209.

Labutina, I. A., Zhivoglyad, A. F., Gorbunov, A. K., Rusanov, G. M., Baldina, E. A., and de Leeuw, J. 1995. The Astrakhanskiy Biosphere Reserve GIS. Part 3: Vegetation map. *ITC Journal,* 3:197–201.

Lychagin, M. Y., Baldina, E. A., Gorbunov, A. K., Labutina, I. A., de Leeuw, J., Kasimov, N. S., Gennadiyev, A. N., Krivonosov, G. A., and Kronenberg, S. B. 1995. The Astrakhanskiy Biosphere Reserve GIS. Part 1: Present status and perspectives. *ITC Journal,* 3:189–192.

Rusanov, G. M. 1993. Status of natural wetlands of the lower reaches of the Volga delta and prodelta under rising Caspian sea level conditions (according to space mapping data) [original in Russian]. In *Osobo okhranyaemye prirodnye territorii basseyna Volgi (Specially Protected Natural Territories of the Volga Basin).* Astrakhan, pp. 29–35.

Saraf, A. K., and Choudhury, P. R. 1998. Integrated remote sensing and GIS for groundwater exploration and identification of artificial research sites. *International Journal of Remote Sensing,* 19:1825–1842.

Smart, M., ed. 1976. *Proceedings, International Conference on the Conservation of Wetlands and Waterfowl,* Heiligenhafen, Germany, December 2–6, 1974. Slimbridge, Gloucester, England: International Waterfowl Research Bureau, 492 pp.

Soeters, R., Rendgers, N., and Van Westen, C. J. 1991. Remote sensing and GIS as applied to mountain hazard analysis and environmental monitoring. *Proceedings, 8th Thematic Conference on Geological Remote Sensing: Exploration, Engineering, and Environment,* Denver, CO, April 19–May 2. Ann Arbor, MI: Environmental Research Institute of Michigan, pp. 1389–1402.

United Nations Environmental Programme. 1993. *United Nations List of National Parks and Protected Areas.* Available on-line via the World Conservation Monitoring Centre, Protected areas information, at: *http://www.wcmc.org.uk/protected_areas/data* (viewed February 8, 1999)

Chapter

14

Changes in Coastal Vegetation in the Northern Caspian Region During Sea-Level Rise

Valentina I. Kravtsova and Elena G. Myalo

Geography Department
Moscow State University
Moscow, Russia

ABSTRACT

A series of satellite images taken from 1975 to 1994 provides the opportunity to investigate the impact of sea-level rise and trends of change in the terrain and vegetation of the coastal zone of the northern Caspian Sea. Over this 20-year period, the 2-m rise in sea level has caused noticeable changes: flooding of the shores, erosion of beach ridges, and a decrease in areas of submergent vegetation. Dynamic elements of coastal vegetation consist of alternating strips of reed, annual saline vegetation, and meadow–solonchak vegetation, with tamarisk (*Tamarix ramosissima*) increasing landward. New vegetation strips have grown up when the old ones die off. Thus, the transgression of the sea has created a return to initial stages of ecological succession in the coastal plain.

14.1 INTRODUCTION

The vegetation zones of the Caspian Sea coast are dynamic and shift in response to sea-level fluctuations. Research into coastal ecosystems and their changes was made urgent by a rapid rise in the level of the Caspian Sea beginning in 1978. Such research is necessary in order to forecast wider changes in natural complexes under various scenarios of sea-level fluctuation (Kaplin and Ignatov, 1997). The study of vegetation, one of the most reactive components of the coastal zone, is especially important, as animal populations, human living conditions, and agricultural activity are all closely tied to the vegetation. The basic data for tracking regional ecosystem dynamics, especially changes in vegetation, come from space photographs.

A series of images taken from the Salyut 4 (1975) and Salyut 7 (1982) orbital stations, the *Kosmos* (1984, 1986) and Landsat (1992) satellites, and the Space Shuttle (1994) allowed us to track changes in the coastal zone terrain and vegetation along the northern Caspian coast during the sea-level rise.

A portion of the northern Caspian Sea coast between the eastern edge of the Volga delta and the present mouth of the Ural River, which was used as an example for our research, is distinguished by a unique scalloped configuration (Figure 14.1) (see also Figures 9.2 and Figure 10.2A). Space photographs of the northern Caspian reveal outlines of the late Pleistocene bay and the ancient Ural delta, which may once have extended to the position of the present Volga delta and occupied a comparable area (more than 12,000 km^2). The scalloped pattern of the coast (the alternation of narrow hollows and ridges, which, when flooded, form bays and peninsulas) is part of the western edge of the Ural delta, created by its former channels and islands. This part of the Caspian coast has undergone recent significant changes due to the sea-level rise, which are clearly shown in the 1975–1994 time series of satellite images (Figures 14.2 to 14.4).

14.2 RESULTS

14.2.1 Coastal Zone Vegetation During the Regression

Figure 14.2 presents four images of the same section of coast (see Figure 14.1 for location), taken in 1975, 1982, 1984, and 1986. In 1975 (Figure 14.2), near the end of the Caspian regression, the coastline configuration was marked by a narrow belt of reeds [editor's note: probably *Phragmites australis*], distinguished in the photograph by an almost black color. Barrier beaches parallel to the coast, generally at a distance of 0.5 to 1 km from the coast, and devoid of vegetation were observed; they form light bands parallel to the coast. In the eastern part of this area, a second beach was observed at a distance of 2 to 3 km from the coast, and still farther to the east, a third beach. They are separated by hollows with submergent vegetation, which creates the typical banded pattern in the photograph.

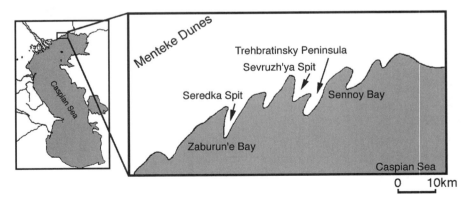

Figure 14.1 Area of investigation, northern coast of the Caspian Sea.

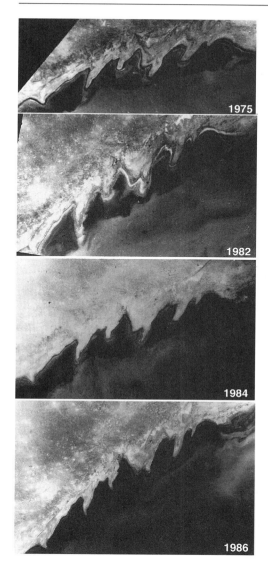

Figure 14.2 Portion of the northern Caspian coastal zone between the Volga and Ural deltas in satellite imagery: from the *Salyut 4* orbital station, 1975; from the *Salyut 7* orbital station, 1982; from the *Kosmos* satellite, 1984; from the *Kosmos* satellite, 1986.

Dark gray, almost black bands of submergent vegetation, which extend for a distance of 10 to 15 km from the coast, are clearly visible behind the band of barrier beaches parallel to the coast on the seaward side. On the seaward side, the submergent vegetation is outlined fairly clearly by an even line. The nonconformance of these bands with the bottom terrain, in which the underwater continuations of shoals and the hollows between them were observed using medium-scale bathymetric maps (with minor depth fluctuations of 1 to 2 m), suggests that the seaward vegetation boundary is dependent on the effects of the inflow current parallel to the coast. This current has been noted in the literature (Leontiev et al., 1977). Submergent vegetation covers the calm zone bounded by this flow. It is apparent that these are thickets of red algae (*Ceramium diaphanum*), a large amount of which Kireeva and Shchapova (1957) noted in the northern Caspian. We also observed thickets of claret-colored algae here during aerovisual observations in 1977.

On land immediately adjacent to the coastline, there are groupings of glasswort (*Salicornia europaea*) with some annual saltwort and seablite (*Salsola* sp. and *Suaeda* sp.) mixed with patches of reed. There are characteristic groupings of annual saltwort and salt-meadow grasses (*Puccinelia dolicholepis, Aeluropus littoralis*) mixed with tamarisk (*Tamarix ramosissima*). In belts a bit farther from the sea, which flood only during periods of great wind surges, similar assemblages are forming (Nikitin, 1954). These dominantly dry areas are light-toned in the images. A strip of solonchaks (mud or salt flats) run along the coast in the narrow hollows that parallel the coast and were also mostly dry during this period except for some places—wet areas of the solonchaks that had singular occurrences of several perennial saltworts (*Kalidium* sp., *Halostachys* sp).

The Menteke sand dunes in the northern Caspian area are separated from the Ural delta by a narrow belt of Baer mounds (ancient parallel ridges of complex origin, probably eolian) that alternate with swales of varying depths, which were dry at the time of mapping in 1975. At the same time, the lakes (locally termed *ilmeni*) between the Baer mounds, were filled with water (the photograph was taken in June 1975) during Volga River flooding. Typical desert vegetation grows on the dunes, predominately "biyurguna" [*Anabasis salsa* (Chenopodiaceae)] and white wormwood (*Artemisia lerchiana*). Slopes and flat intermound areas are covered with salt-meadow vegetation, and the dry beds at the bottom of the depressions are covered with solonchaks and *Halocnemum strobilaceum* around the periphery (Shvyryaeva, 1961).

14.2.2 Vegetative Responses to the 1982–1986 Sea-Level Rise

In 1982 the sea level had risen 0.87 m above its 1977 lowstand (Table 14.1). Compared to 1975, the 1982 coastline was also marked by a solid belt of reeds (a gray tone in the image rather than a black tone, as the image was captured in late summer when the reeds were tall, Figure 14.2). When the sea level rose, this strip encroached landward (by 0.5 to 1.5 km) into the areas that were wet solonchaks in 1975. Despite the almost 1-m rise in sea level, the coastline configuration essentially did not change, except for the upper part of Sennoy Bay, where the shoreline moved 3.5 km inland. Moreover, the coastline on the sea side began to be edged with light belts that correspond to barrier beaches, formation of which was activated by increased depth and strong wave action. Two island ridges that formed on the end of Seredka Spit testify to this.

TABLE 14.1 Caspian Sea-Level Benchmarks Cited in This Chapter

Year	Sea Level (m)
1977 (lowstand)	−29.02
1984	−28.08
1986	−27.91
1992	−27.05
1994	−26.70

The belt of submergent vegetation, represented by a dark-gray tone in the image, grew inland in places and covered all the bays right up to the barrier beaches. On the sea side the vegetation is delineated by a straight line, possibly due to the effects of inflow currents of wind-surge waters. On the land-side coastal plain, we did not observe notable changes in flooding and vegetation coverage. We could not identify changes in the belt of Baer mounds separating the delta from the Caspian desert, and the intermound swales were not flooded, despite the almost 1-m rise in the sea level. It is possible the lack of flooding was due to the late summer image period.

In 1984 the sea level was 0.94 m above the 1977 level (Table 14.1) and the images recorded a similar situation (Figure 14.2). The coastline configuration had hardly changed. The reed belt bordering the coastline on the land-side coastal plain was very narrow, less than 1 km, but on the sea side reeds grew almost everywhere on the barrier beaches that had appeared. The thickets of submergent vegetation in the bays, which extended along the entire length of the coast, appeared as a 10- to 15-km-wide belt, the configuration of which was almost unchanged. The swales between the Baer mounds that form the western edge of the Ural delta remained dry. When sea level rose by 0.94 m, it was undoubtedly accompanied by a rise in the groundwater table and changes in soil salinity. However, the vegetation did not undergo any substantive changes, and this apparently is explained by the wide ecological range of the dominant vegetation and by the youthfulness of the flooded coastal plain.

We observed a similar pattern in 1986 when sea level had risen 1.11 m from 1977 (Table 14.1, Figure 14.2). The coastline configuration remained the same almost everywhere, except for the Trekhbratinsky Peninsula, where a 3-km-wide strip of land was flooded. The narrow belt of reed thickets continued to line the coast, but in the lowest portion of the belt where the land was flooding (the western coast of the Sevruzh'ya Spit), this belt was now inundated by the sea, 1 to 2 km from the coast.

As before, submergent vegetation in the shallow zone covered a wide belt stretching 10 to 15 km from the coast. We did not note changes in the vegetation covering of the new coastal plain. The light spots of the solonchaks in the longshore belt are clearer than in previous images; the solonchaks attach to the Primorskij canal, which leads out of the Volga lakes from the west and out of the Ural branches from the east and is just identifiable. The belt of Baer mounds remains dry and the light spots of the solonchaks are more clearly outlined in the intermound swales. It is possible that the soils have become salinized by the rise in the groundwater level following the rise in the sea level.

14.2.3 Vegetation Changes in the 1990s

The situation changed sharply from 1991 to 1992. We analyzed a Landsat satellite image taken in 1992 when sea level was 1.97 m above the 1977 level (Figure 14.3); The shoreline moved inland 3 to 15 km and continued to maintain a scalloped configuration due to the terrain—the hollows were flooded and became bays, and the ridges became peninsulas.

The characteristic barrier beaches, which are marked by the narrow belt of reeds (rust-colored in Figure 14.3), have a scalloped form and are the same barrier beaches

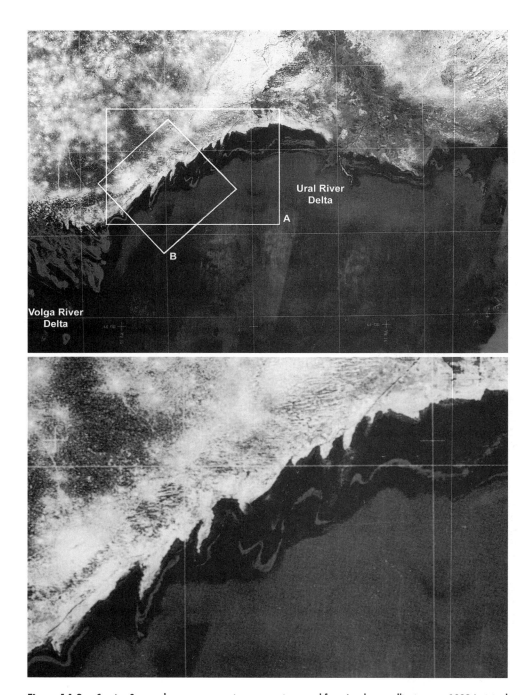

Figure 14.3 Caspian Sea northern coast on an image mosaic created from Landsat satellite images, 1992 (original in color). Box A denotes the area shown in greater detail in the lower half of the figure. Box B denotes the area of the Space Shuttle photograph STS064-101-25 shown in Figure 11.4.

whose formation was observed in the 1982 photographs. The previous wide (10 to 15 km) belt of underwater vegetation extending along the coast has now narrowed and is only 3 to 5 km wide. With increased depth, the belt of aquatic vegetation (algae and pond weeds) shifted to cover areas where annual saltwort vegetation had died out.

Areas of groundwater rise in the belt of solonchaks on the coast are clearly visible. The areas are deep blue on the false-color photograph. Salt-meadow vegetation is forming here with significant amounts of perennial saltworts (galophytes *Halocnemum strobilaceum* and *Kalidium foliatum*). The belt of Baer mounds and intermound swales that edged the delta from the west have also been flooded, and a significant portion of the swales are under water. Groups of reeds and annual saltworts such as jointed glasswort (*Salicornia herbacea*) developing in the outlying areas show as red on the Landsat images. The desert vegetation on the dry portions of the Baer mounds has been preserved only in narrow bands on the crests of the mounds.

The worst coastal flooding was registered in a series of detailed large-scale images taken in 1994 from the Space Shuttle when the sea level was 2.32 m above the 1977 lowstand (Table 14.1, Figure 14.4). Reed thickets are all that remain of beaches on the inundated 1982–1984 coastline. The dark contours of water with submergent vegetation in 1994 show that the beach had been only partially preserved (Figure 14.4). On the sea side, these contours continued to erode. The flooded coast, covered with saltwort, has a light appearance; it is possible that the shallow zone was free of vegetation by 1994 except for the coastal band, where the darker tone of the image attests to its presence. On the image it is easy to observe the separate tracts of reeds, which apparently, are attached to the elevated portions of the bottom. This belt of coastal vegetation makes it difficult to identify the coastal contours clearly.

The transgression of the coastline had by this time reached the belt edging the Baer mounds and the intermound swales that are remnants of the ancient delta. In the northern part they are under water, but the flooding is not as noticeable in the

Figure 14.4 Photomosaic of the northern Caspian shore, extending from the study area toward the Ural delta, September 1994 (NASA photographs STS064-101-25, 26, and 27). The former coastline appears as bands of vegetation offshore; the shape of the new coastline close to the Ural is dominated by dams and levees.

eastern part, possibly because the photograph from 1994 (Figure 14.4) was taken during a drier season than was the image in 1992 (Figure 14.3, when the flooding of the swales and the formation of lakes was observed everywhere). Apparently, wetting in the nonflooded swales increased and groups of annual saltworts developed in them. Groups of reeds may also have grown in the swales as evidenced on the images by a darker color. Farther east, the coastal outline in the present area of the Ural delta became linear during the sea-level rise, due to the presence of dams which create the anthropogenic character of the coastline (Figure 14.4).

14.2.4 Succession of Vegetation

Analysis of a time series of images allows us to observe the dynamics of vegetative cover in connection with sea-level rise. For the coastal plain as a whole, the characteristic change in the vegetation belt has been the shift inland from the coastline of the reed belt, the belt of annual saltwort vegetation with isolated patches of reeds, and the belt of annual saltwort and salt-meadow vegetation with tamarisk groves. This change in the belt, to a large degree, demonstrates the return to an early successional stage in the new coastal plain. In areas where the Baer mounds are developed, desert vegetation grows on the tops of the mounds, while salt-meadow vegetation or solonchaks are found in the swales.

The landward shift of the vegetation belt is the most important aspect of vegetation dynamics connected with sea-level rise and the landward movement of the coastline. All basic types of plant communities have been preserved; however, several changes in their structure have been noted on the surface of the coastal plain. For example, the belt of submergent vegetation contracted, shifting toward the land because of the increase in depth, but it still does not extend into the recently flooded areas.

During the regression period in the mid-1970s, a relatively narrow but fairly well formed belt of reeds was observed. At present the singular relics of the belt are isolated tracts of reeds extending along the coast, preserving the old coastal configuration; they are undoubtedly attached to the highest portion of the bottom. At the same time, the first formative stages of a reed border, also represented by isolated tracts, have been noted along the new coastline.

In the new coastal plain along the 1994 coastline, annual saltworts are growing and the reeds, salt meadow, and tamarisk are moving farther inland. In this way, it is as if there is a reversion to the initial stages of ecological succession along the coastal plain. In the belt of Baer mounds located far from the sea, changes in vegetation are observed only in the swales, due to the regional rise in the groundwater level.

14.3 CONCLUSION

Analysis of a time series of satellite images shows that over the course of the first 10 years of the transgression, a sea-level rise of approximately 1 m possibly cushioned the effects of the Volga waters on the given portion of the coast, and significant changes in the ecological condition of the coast did not occur. The coastal zone terrain in general remained as it was previously; only several of its borders were observed to change by forming sedimentary barrier beaches due to strong wave action. Significant changes in the vegetation during this time were not noted in the satellite images.

The next 10-year interval and sea-level rise of another 1 m produced significant ecological changes. These changes included coastal flooding, beginning with the preservation of the barrier beaches and followed by erosion of the beaches, of which only fragments finally remained. Submergent vegetation on the sea side of the beaches decreased. A change in ecological conditions on the coast was evident from the deep inland incursion of coastal vegetation belts. This incursion has taken place gradually and isolated, dying fragments of vegetation belts are preserved concurrently as new bands are formed. The vegetation of the newly developing belt is returning to earlier successional stages, and the coastal territory as a whole is undergoing ecological succession.

The data obtained provide an idea about the basic conformity of coastal succession during a sea-level rise and the rate of development of individual stages; these data may be used in ecological monitoring of the Caspian coastal zone. Research on vegetation dynamics—the step-by-step changes following sea-level rise—is not only of significance to the Caspian region. This research may be considered as a model in connection with the fluctuations now being observed in other regions, particularly those expected in the near future on a global scale.

ACKNOWLEDGMENTS

The authors acknowledge the efforts of TechTrans International, Inc. (Houston) in translating the manuscript into English, C. A. Evans and J. A. Robinson for assistance with figures, and V. Klemas and staff members of the JSC Office of Earth Sciences for constructive reviews of the paper.

REFERENCES

Kaplin, P. A., and Ignatov, E. I., eds. 1997. Geoehkologicheskie izmeneniya pri kolebaniyakh urovnya Kaspiiskogo morya (Geoecological changes during Caspian sea-level fluctuations). In *Geoecology of the Caspian Region*, Vol. 1. Moscow, 201 pp.

Kireeva, M. S., and Shchapova, T. F. 1957. Materialy po sistematicheskomu sostavu i biomasse vodoroslei i vysshei vodnoi rastitel'nosti Kaspiiskogo morya (Systematic composition and biomass data on algae and higher aquatic vegetation of the Caspian Sea). *USSR Academy of Sciences Oceanic Institute Works*, 23:125–137.

Leontiev, O. K., Maev, E. G., and Rychagov G. I. 1977. *Geomorfologiya beregov i dna Kaspiiskogo morya (Geomorphology of the Caspian Sea Coasts and Bottom)*. Moscow, 208 pp.

Nikitin, S. A. 1954. Rastitel'nost' mezhdurechii Kushuma, Urala i Ehmby i ee kormovye resursy (Vegetation of the area between the Kushum, Ural, and Ehmba rivers and its feed resources). In *Voprosy uluchsheniya kormovoj bazy v stepnoj, polupustynnoj I pustynnoj zonakh SSSR (Questions on Improving the Feed Base in the Steppe, Semi-arid, and Arid Zones of the USSR)*. Moscow–Leningrad, pp. 58–86.

Shvyryaeva, A. M. 1961. Ispol'zovanie rezul'tatov geobotanicheskogo deshifrirovaniya aehrosnimkov pri landshaftnykh obsledovaniyakh v Severnom Prikaspii (Use of geo-botanical interpretations of aerial imaging during landscape surveys of the northern Caspian region). In Miroshnichenko, V. P., ed. *Primenenie aehrometodov v landshaftnykh issledovaniyakh (Application of Aerial Methods in Landscape Research)*. Moscow–Leningrad: USSR Academy of Sciences, pp. 191–216.

Chapter

15

Dynamics of the Northeastern Caspian Sea Coastal Zone in Connection with Sea-Level Rise

Valentina I. Kravtsova

Department of Geography
Moscow State University
Moscow, Russia

ABSTRACT

The northeastern Caspian Sea is distinguished by a large shallow area. We mapped bottom topography and sediments, underwater vegetation, and the subsea landscapes using photographs taken during the regressive stage in 1976. At that time ridge-and-swale topography formed in response to wave action, wind-induced surge phenomena, and currents. Data from subsequent space photographs from 1985, 1992, and 1996 showed that in the transgressive stage, passive flooding of the land occurred without noticeable changes to the coastal zone profile. We documented extensive flooding of dry land and solonchaks and dry hollows that were saturated with water. Ridge-and-swale topography may be subject to future erosion in response to rising sea level.

15.1 INTRODUCTION

The northeastern Caspian, which occupies the area between the mouth of the Ural River and Buzachi Peninsula, is a unique expansive shallow basin with an area of approximately 100×150 km (Figure 15.1; see the color insert) (See also location map and photographs (Figures 9.1, 9.2, and 10.4). It is encircled by coastal mud flats with widely developed wind-induced surge features. The principal distinction of this region of the Caspian is its shallowness. During the regressive stage in the late 1970s, depths in most of the 50-km-wide coastal band that we studied were just 2 to 3 m, and only in the southwestern part (in the Ural canal) did the depth reach 5 to 6 m.

With such shallow depths, fluctuations in sea level change the situation over an extensive surface area, causing a change in sea-bottom relief-forming processes and a change in the wetness of the surrounding coastal plains. For example, on the northwestern coasts of the Caspian in the Kalmyk region, which is also characterized by mudflats and wind-induced surge features, the rise in the level rearranged the coastal zone profile, forming coastal banks and lagoons (Kravtsova and Lukyanova, 1997a, b). There is also an interest in tracing changes in the coastal flats, such as the northeastern Caspian coast, viewing these changes and coastal zone responses as elements of a model for possible future changes during fluctuations in the level of the world oceans.

In 1976, when the sea level approached the minimum, this portion of the Caspian Sea was photographed during the Soviet–German "Raduga" experiment, involving multizone mapping using the MKF-6 camera created jointly by Russian and German specialists and manufactured by the Karl Zeiss company. Comparison of this image with more recent ones may provide answers to the questions posed.

15.2 RESULTS

15.2.1 Regression Stage, 1976

The September 1976 photograph, with exceptional water clarity, recorded bottom relief, bottom sediments, and submerged vegetation in the expansive shallow portion of the Caspian Sea (Figure 15.2). Synthesized natural-color photographs enabled thematic and complex interpretation of coastal area and sea-bottom objects. Interpretations of space photographs were supplemented with field data in 1977—visual observations were conducted from a helicopter, and soil and bottom vegetation samples were taken from aboard a ship. A series of thematic maps were created of this portion of the Caspian coastal zone at a scale 1:500,000 (Kravtsova and Ion, 1982). The map series included bathymetric, geomorphological (Kravtsova et al., 1980), bottom sediment (Ignatov et al., 1980), underwater vegetation (Kravtsova and Ponomareva, 1980), and underwater landscapes (Antonova et al., 1980). The maps show the development of unique bottom relief forms and underwater landscapes. Figure 15.1 is our geomorphological map compiled from the data. It serves as a basis from which to compare the changing coastal zone as sea level rose from 1977 to the present. The relief of the shallow coastal area—coastal swaths about 50 km wide—was formed by the effects of a complex combination of several factors: wave action, wind-induced surge currents, and circular offshore currents. As a result, distinctive ridge-and-swale topography formed here (the local name for the troughs is *shaligy*), which was captured in detail on the map (Figure 15.1).

In the regression stage in the deeper waters of the southwestern region, the bottom was a flat, slightly undulating sedimentary plane that was formed by wave processes, with bottom irregularities created by submerged relief of subaerial genesis. Figure 15.2 shows some of these features. The bright swath of the Agrakhano-Emba fault (1) and the dark channels of the proto-Emba and proto-Ural riverbeds (2), which formed during the more extensive Pleistocene regression of the Caspian Sea, are clear. Sandy terrigenous sediments with bottom vegetation, which are indicators of these channels, are related to the lowstand.

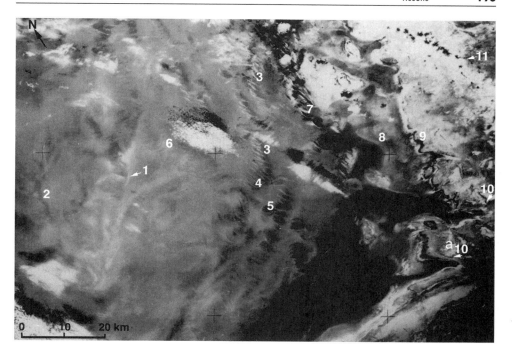

Figure 15.2 Northeastern Caspian Sea photographed from Soyuz 22 in September 1976. The numbered features are discussed in the text. The letter *a* refers to a feature compared across photographs in the text.

The crests of the underwater ridges, which rise above the sea surface in the form of narrow islands (3) or are observable under the water close to the water surface (4), are clearly visible in the photograph. Formed of shell deposits, they have a natural orange color and are roughly the same color (Figure 15.2). They separate the interridge depressions filled with bottom vegetation (5), which are greenish brown in the image. The crests and troughs form as many as five chains, which generally run parallel to the coast. Series of crests and troughs in the initial stage of formation (6) are observed in the deeper-water section at depths of 3 to 5 m. The troughs are most pronounced in the shallow zone at depths of 1.5 to 3 m, while relict emergent crests in the form of narrow islands (7) are observed in coastal areas with depths of up to 1.5 m.

Several former Caspian shorelines, partly covered with saltwater vegetation, can be detected in the coastal zone from landscape features. On Figure 15.2 they are labeled as follows: (8) the flood line of the shore during the wind-induced surges of 1976, (9) the 1973 shoreline that corresponded to a sea level of −28.7 m, (10) the 1940 shoreline at −27.8 m, and (11) the 1929 shoreline at −26 m (Figure 15.3).

15.2.2 Beginning of Transgression, 1985

What happened to the bottom topography of this region after the rise in sea level? To follow-up our original study, we mapped the same region of the Caspian coast

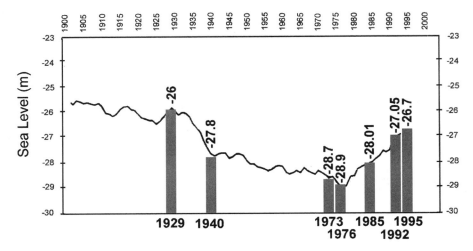

Figure 15.3 Benchmarks of Caspian sea level used for reference in this study. Bars indicate sea level in the indicated year as reported by the authors. Line indicates data for the twentieth century used by Varuschenko et al. (Chapter 16).

using contemporary space mapping materials, tracking features and their transformation as the sea level rose. Photograph STS51F-34-67 (Figure 15.4), taken in 1985 from the Space Shuttle, shows changes in the coastal zone. The level of the Caspian in 1985 was −28.01 m (i.e., it had risen 1.01 m since 1977) (Figure 15.3). Through careful comparison with materials from 1976 (Figures 15.1 and 15.2), one can identify several coastal topographic forms that have been preserved. The coastline has a position slightly higher than that of 1940, shown on the map (Figure 15.1). This is clearly visible from the readily identifiable outline of the island (a, in Figure 15.4), which was connected to land during the lowstand in 1976 (a, Figures 15.1 and 15.2) but had previously been an island in 1940 (Figure 15.1).

The scalloped coastline of 1940 in the western portion of this island can be seen beneath the water. The presence of old shelves on the eastern coast of the island caused the previous configuration to be preserved. The contours of the remaining islands also correspond fairly well to subaerially exposed islands in 1940 marked on Figure 15.1. The former bay-lagoon (b) shown on the map is flooded, and the new coastline passes through the Prorva River estuary, which is drying up.

The Tengiz oil field began to be developed here; on the photograph a large smoke plume is visible at this oil field (dark black plume just beneath the label "Tengiz"). Rising sea level and the threat of flooding of this territory was the wake-up call to build protective dams here; fragments of them are visible in the photograph as light bands (marked with arrows). Certain parts of the ridge-and-swale topography can be traced in the sea with great difficulty, but the ridges could have been eroded when the depth changed. However, bands of muddy water obscured bottom features, and observation conditions at the time of photography were not optimal. The fact that these bands stretch along the coast indicates continuing wind-induced surges.

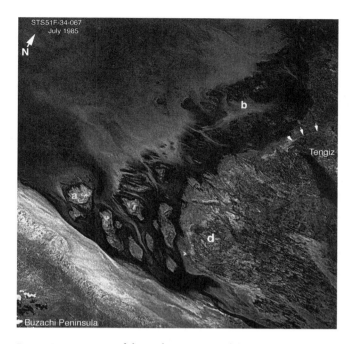

Figure 15.4 Portion of the northeastern coast of the Caspian Sea near the mouth of Komsomolets Bay and the Mertvyy Kultuk dry bed between Prorva and Buzachi Peninsula August 1, 1985, (NASA photograph STS51F-34-67). The letters a, b, and d mark features compared among figures.

15.2.3 Continuing Transgression, 1990s

Photomosaics of 1992 Landsat images (Figure 15.5), show the same part of the Caspian when sea level was at −27.05 m, 1.97 m above the relative lowstand in 1977 (Figure 15.3). Some of the same features noted in Figures 15.1, 15.2, and 15.4 are also marked on Figure 15.5 for reference (islands, a and d; the flooded bay-lagoon, b).

The zone in which wind-induced surge features were forming in 1977 is completely flooded and partially covered by saltwater vegetation (c). In 1992 the Caspian shore should have occupied a position between the 1940 and 1929 shorelines. However, the coastal configuration in the north-trending segment of the northeastern coast does not correspond in any way to the old shoreline configurations, due to anthropogenic changes. Protective structures (dams) built while Tengiz oil field was being developed create rectilinear contours of the coastline in places (marked with arrows near b), in contrast with the old coastline.

Flooding is observed over a wide swath of land (up to 50 km); there are lakes between Baer mounds (ancient parallel ridges of complex origin, probably eolian), and deflation hollows are flooded. The blue east-aligned contours of the lakes emphasize the shapes of the Baer mounds, and the azure spots of deflation hollows

Figure 15.5 Mosaic of the northeastern coast of the Caspian Sea compiled from Landsat images, 1992.

are reminders of the more recent aeolian history. Flooding and excess moisture have complicated agricultural activity here, caused transportation problems, and have created serious difficulties for oil extraction.

The northern coast of Buzachi Peninsula was also badly flooded, and it was necessary to wall off the oil fields with a dam (bottom arrows, Figure 15.5). The expansive areas of dry beds in the central portion of the peninsula turned into wet solonchaks. The upper portion of the Mertvyy Kultuk and Kaydak dry beds were also flooded and turned into salt marshes (wet solonchaks).

On the shallow seafloor, topographic forms that correspond to former beach ridges along the shores of former lagoons can be traced. The hills have been completely covered by water and could not be traced for any distance, possibly because of low water clarity. They may also have been eroded during the transgression and/or submerged below the effects of wind-induced surges. In the sea, small islands and sandbars covered with reedy vegetation have been preserved and are distinguished in the photograph by their red color (southwest of b, Figure 15.5). The islands and bars are parts of relict hills in the coastal zone and ridges on the shore in the zone of wind-induced surge.

In March 1996, the portion of the northeastern Caspian between Buzachi Peninsula and the Mertvyy Kultuk dry beds was recorded in Shuttle–*Mir* photographs NM21-704-12 and NM21-704-15. These photographs were taken when the coastal waters were still covered with ice and provide a clear, high-contrast image of the coastline. Sea level in 1995 was -26.7 m (i.e., 2.32 m above the 1977 level) (Figure 15.3), and further flooding of this very flat part of the coast had occurred. The area covered by the two photographs comprises only the lower right corner of Figure 15.4 or 15.5. Comparing the photographs from 1985 and 1996 (Figures 15.4 and 15.6), it is possible to identify landward movement of the coastline and to recognize those portions that were flooded (compare island d, Figures 15.4 to 15.6). The coastline configuration indicates that this line is unstable and that there is passive inundation of the western portion of the Mertvyy Kultuk by seawater.

The irregularity of the coast and the numerous east-trending peninsulas and bays are results of the presence of Baer mounds. The nature of the coast is different in the eastern part of Mertvyy Kultuk, framed by the disappearing sea cliffs of the 1930s and 1940s. There, the shifting of the capes and coves indicates prevalence in the past of erosional and depositional coastal processes that are now reappearing. The Kaydak dry bed in the 1996 photographs is completely flooded (Figure 15.6), in contrast to 1992, when wet solonchaks were observed there (Figure 15.5).

15.3 CONCLUSION

We studied coastal changes with rise in sea level in the northeastern Caspian using a variety of images taken from space. In this area, distinguished by a very large coastal mud flat, passive flooding of the land took place without noticeable changes to the coastal zone profile and without the formation of specific coastal relief forms such as coastal ridges and lagoons. This stands in contrast to the northwestern shores of the Caspian (Kalmyk coast). Although the northwestern Caspian also includes a coastal mud flat, rise in sea level induced modification of the coastal-zone profile (see Chapter 19).

In the northeastern Caspian, marked flooding of the land and the water saturation of solonchaks and dry beds is noticeable. Ridge-and-swale topography formed in shallow Caspian waters during the regression due to the effects of wave action, wind-induced surges, and currents; those bedforms may now be subject to erosion as water depth increases.

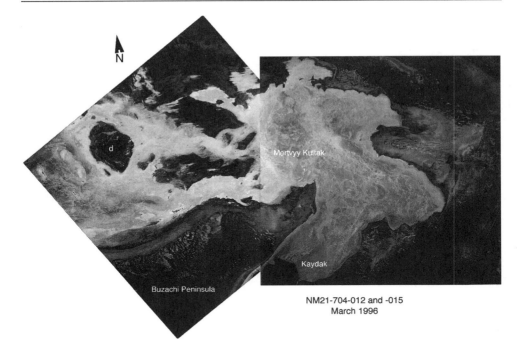

Figure 15.6 Mouth of Komsomolets Bay and the Kaydak dry beds, flooded with water and frozen March 1996 (NASA photographs NM21-704-12 and NM21-704-15). The letter d marks a feature compared among figures.

ACKNOWLEDGMENTS

The author acknowledges the efforts of: C. A. Evans and L. Prejean for assistance with figures, TechTrans International, Inc. (Houston) for translating the manuscript into English, and M. A. Helfert and staff members of the JSC Office of Earth Sciences for constructive reviews.

REFERENCES

Antonova, S. Yu., Kravtsova, V. I., and Ushakova, L. A. 1980. Cartography of underwater landscapes [original in Russian]. In *Kosmicheskaya semka i tematicheskoe kartografirovanie (Space Photography and Thematic Mapping)*. Moscow, pp. 190–198.

Ignatov, E. I., Kravtsova, V. I., Ushakov, L. A., and Shipilova, L. M. 1980. Study and cartography of the ground [original in Russian]. In *Kosmicheskaya semka i tematicheskoe kartografirovanie (Space Photography and Thematic Mapping)*. Moscow, pp. 166–181.

Kravtsova, V. I., and Ion, K.-H., eds. 1982. *Deshifrirovanie mnogozonal'nykh aehrokosmicheskikh snimkov: Metodika i rezul'taty (Interpreting Multizone Aerospace Photographs: Technique and Results)*. Moscow: Science; Berlin: Academy Verlag, pp. 6–12.

Kravtsova, V. I., and Lukyanova, S. A. 1995. Study of the dynamics of the Caspian Kalmyk coast using multitemporal space images [original in Russian]. *Moscow State University Bulletin, Geography Series,* 5:51–58.

Kravtsova, V. I., and Lukyanova, S. A. 1997a. Transgressive changes in the Caspian Sea Russian coastal zone (based on the interpretation of aerospace photographs) [original in Russian]. *Geomorfologiya (Geomorphology),* 2:35–45.

Kravtsova, V. I., and Lukyanova, S. A. 1997b. Mapping of the dynamics of the Caspian Sea coastal zone by multitemporal space images. *Proceedings, 18th ICA/ACI International Cartographic Conference ICC97,* Stockholm, Sweden, June 23–27, Vol. 1, pp. 67–73.

Kravtsova, V. I., and Ponomareva, E. D. 1980. Cartography of underwater vegetation [original in Russian]. In *Kosmicheskaya semka i tematicheskoe kartografirovanie (Space Photography and Thematic Mapping).* Moscow, pp. 181–190,

Kravtsova, V. I., Leontiev, O. K., and Romanenko, L. S. 1980. Study and cartography of bottom relief [original in Russian]. In *Kosmicheskaya semka i tematicheskoe kartografirovanie (Space Photography and Thematic Mapping).* Moscow, pp. 149–166.

Chapter 16

Evolution of the Gulf of Kara-Bogaz-Gol in the Past Century

A. N. Varushchenko, S. A. Lukyanova, G. D. Solovieva, A. N. Kosarev, and A. V. Kurayev

Department of Geography
Moscow State University
Moscow, Russia

ABSTRACT

Using a variety of sources—historic maps, photographs from Russian satellites and spacecraft, and photographs from the U.S. Space Shuttle—we document the history of changes in the Kara-Bogaz-Gol basin during the twentieth century. Changes in the water level are linked to regression and transgression in the main basin of the Caspian Sea. Because the Gulf of Kara-Bogaz-Gol is so shallow, the changes are even more dramatic than those observed along other Caspian shorelines and range from complete drying to filling of the basin. Human intervention by damming and controlling flow through the Kara-Bogaz Inlet increased the magnitude of changes observed and influenced salinity and water chemistry. We link the patterns of coastal change observed to the interaction between rapidly changing hydrology and the underlying geology. Key processes observed include evolution of peninsulas and islands, delta formation at the inflow, and the effects of bottom smoothing and winds on water movement within the basin.

16.1 INTRODUCTION

The largest lagoon of the Caspian Sea, the Gulf of Kara-Bogaz-Gol (see also Figures 9.3 and 10.5), dates back to an open bay in at least the Upper Pliocene. Remnants of later-eroded marine depositional features, developed along the edges of this bay, now reach the surface at some points, especially in the region of Bekdash (Leontyev, 1961). Marine depositional features of the barrier type were also formed here in the early Pleistocene, as indicated by the outcrops of Baku (lower Pleistocene) shell limestones, chiefly along the southern coast of the gulf. The gulf only became a lagoon in the Late Khazar period (end of the middle Pleistocene), when the mouth portion of the wide bay was blocked by a large barrier growing from the north southward from the ledge of the Kenderly-Kayasan Plateau in the region of Bekdash; initially, it was formed as a coastal barrier (Leontyev, 1961). Relics of this enormous barrier, in the form of numerous shell limestone outcrops, may be found in the body of the now-existing Holocene sand barrier, as well as under water on both sides of it. After isolation from the sea, Kara-Bogaz-Gol was connected to it only through a narrow inlet which divided the Holocene depositional barrier into two parts (North and South Kara-Bogaz Spits). The inlet parameters have changed over time depending on the correlation of the water level in the gulf and in the sea. Intense evaporation from the gulf surface causes steady inflow of Caspian waters into it. Hence the name of the gulf: *Kara-bogaz,* translated from the Turkish, means "black jaws," which steadily and irretrievably gulp down the seawater.

As seawater influx is the one source feeding this expansive lagoon, its entire history is most closely associated with the sequence and scope of Caspian sea-level fluctuation (Figure 16.1). This is evident in maps of the most recent stages of this history (e.g., Figure 16.2), as well as from aerial and space photographs. The Kara-Bogaz Inlet and debris cones from it are visible in space photographs. Analysis of cartographic and photographic materials from different years reveals evolution of morphological features in the Gulf of Kara-Bogaz-Gol throughout the century.

16.2 NATURAL CONDITIONS

The Kara-Bogaz-Gol region is arid, with a hot summer (maximum air temperature reaching 40°C; average annual temperature around 13°C) and strong winds in autumn and winter. Average precipitation is 75 to 115 mm/yr; evaporation is 975 to 1000 mm/yr. As a consequence of the intensive evaporation from the gulf surface, its waters are extremely saline, higher than 200‰.

The physical and geographic conditions of Kara-Bogaz-Gol, as well as its hydrological and hydrochemical regime, which depend on the Caspian level and on inflow into the gulf, have experienced marked long-term variations. When there is signifi-

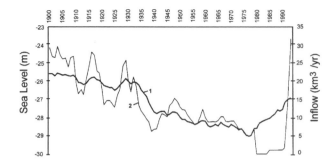

Figure 16.1 Long-term changes in the level of the Caspian Sea (1) and water inflow into the Gulf of Kara-Bogaz-Gol (2).

Figure 16.2 Diagram of the Gulf of Kara-Bogaz-Gol: 1, shoreline of sea and bay per 1930 maps; 2, schematic boundary of water's edge in 1956; 3, exposed part of upper layer of salts in windless weather; 4, gypsum salt flats (solonchaks). (Adapted from Dzens-Litovskiy, 1959.)

cant filling of the gulf with water, the mass of inflowing seawater is split at the inner mouth of the inlet into two principal currents. One runs northward along the North Kara-Bogaz Spit, gradually turning northeast; the other flows southward.

These currents develop only in the western part of the gulf. In other areas the currents are determined by the wind over the waters. Wind-induced currents in the gulf are usually on the order of 10 to 30 cm/s. The maximum recorded speed was 65 cm/s, although currents may be faster with strong winds. Intensified dynamic activity of the waters may be observed in the western part of the gulf, especially at the mouth of the inlet, where there is pervasive mixing of the seawaters with the gulf brine, as is evident in photo STS068-247-5 (Figure 16.7) in the form of eddies and other irregularities.

16.3 GEOMORPHOLOGICAL CHANGES

Early in this century (before 1929), when Caspian sea level fluctuated around −26 m, the Kara-Bogaz-Gol basin was almost entirely filled with water. The area of its water surface was a little more than 18,000 km^2, and its maximal depth was 9 m (Leontyev, 1961; Terziyev et al., 1986). Inlet discharge in these years reached its greatest value, 18 to 25.8 km^3, and the length of the inlet reached 6.8 km (Mironova, 1959). The gulf and sea-level difference was minimal and in 1921 amounted to just 0.44 m.

Marine erosion was the main process of coast development, especially along the irregular northern and eastern coastal bluffs. At the foot of the latter there were only small, temporary pebble and shell beaches. Stable depositional features were preserved only at a few capes formed by proluvial fans from large ravines (e.g., Kulan-Gurlan). Erosional processes were typical at the time also for the lower (western and southern) coasts of the gulf. In the south there was active erosion of the limestone ends of the tombolo Omchali Peninsula and similar peninsulas, which were remnants of ancient marine depositional features of Baku (lower Pleistocene) age, and in places (in the southeast), perhaps outcrops of Cretaceous rocks connected to the coast by Khvalynian (upper Pleistocene) and Holocene sand bulkheads. Between the peninsulas and the proximal part of the South Kara-Bogaz Spit, the processes of long-term erosion led to smoothing of the low sea cliffs, which bounded the New Caspian (Holocene) and Late Khvalynian (upper Pleistocene) marine sedimentary terraces. The smoothed scarp of the gulf's bedrock coast east of Omchali Peninsula was subjected to active erosion. The eastern edge of the sandy North Kara-Bogaz Spit was also eroded, reflected by the sea cliff several kilometers north of Kara-Bogaz Inlet.

An abrupt 1.8-m lowering of Caspian level in the period 1929 to 1940 and a number of subsequent episodes of lesser decline have reduced water supply to the gulf. In 1948 the Caspian level was –27.87 m. Caspian water inflow into the gulf from 1948 to 1952 decreased to 12 to 14 km^3, and in 1956 to 8.0 km^3 (Mironova, 1959). The difference in the gulf and Caspian sea levels increased to 3.17 m in 1947 and 3.80 m in 1955. By 1957–1959 the gulf water surface area was no more than 13,000 km^2, and its maximum depth was less than 3 m. According to Leontyev (1961, p. 35), "the Kara-Bogaz Inlet has been narrowed and extended, becoming the only 'sea river' in the world with a rapid current, a swiftly growing delta and even a waterfall...". The inlet narrowed at the intersection of the relict Pleistocene (Late Khazar) barrier formed by lithified coquina. In 1951 the height of the waterfall was 1.6 m and the inlet was 9 km long and 120 to 300 m wide (Mironova, 1959). Although it was still accessible to small vessels in 1933, the inlet lost its navigational importance in the 1950s.

Despite the drop in Caspian water level, no marked shoreline progradation on the seaward side of the depositional barrier was noticed from 1957 to 1958. A narrow sand and shell beach ran along the steep western edge of the barrier. On the eastern side of the barrier, conversely, significant changes occurred because of the faster drop in water level. The chief effect of this was the almost universal abandonment of sea cliffs and drying of extensive areas of the bottom. In comparison with the position of the shoreline in 1940 maps (Leontyev, 1961), it is clear that this drying has been most evident near the southern and eastern ends of the gulf (Figure 16.2). By 1957, the southeast corner of the gulf was completely dry; in the east the shoreline had retreated 30 to 40 km, and only at the depositional Cape Kulan-Gurlan was it somewhat less (8 km). Many coastal islands (e.g., Taraba and Secstan) had lost their island status and become linked to the shore, owing to the drying up of the bay bottom sections that separated them. In the southwest, between the Omchali Peninsula and South Kara-Bogaz Spit, the shoreline had retreated northward 10 to 12 km. However, at the end of the Omchali Peninsula, the shoreline retreat from the foot of the sea cliff formed of Baku (lower Pleistocene) shell limestones was no more than 0.3 to 0.4 km.

During the retreat of the shoreline, a broad strip of former bay bottom was exposed, a flat plain that had gradually been submerged. It is covered with a salt crust, the drying of which provokes the appearance of various microrelief features: warping, polygonal cracks, distension, and so on. In the 1960s and 1970s the gradual drop in the Caspian sea level continued. For instance, in 1957 it was –28.47 m, and in 1957–1977, –29.02 m. The water level in the gulf was 4.5 m lower than the Caspian sea level, the inlet length increased to 11 km, and the height of its threshold waterfall increased to 4 m.

Small-scale space photos of these years, obtained from the Soyuz 9 spacecraft and the Meteor 25 satellite, show a consistent reduction in the water area of Kara-Bogaz-Gol. By 1970 (Figure 16.3), it had decreased to roughly 8400 km^2. Drying of the bay bottom reached its greatest extent along its northern, northeastern, and eastern sides. In the northeast, the smoothly curved line of the retreating water boundary intersected the gulf basin diagonally near its center, running in the direction of Cape Kulan-Gurlan, and then proceeding southward to the distal end of the Jangy-Su Spit. The dried bottom plain, covered with a salt crust, may be viewed in the space photo. The sharp salt ledge south-southwest of Cape Kulan-Gurlan is conspicuous. It is probably associated with outcrops of Cretaceous bedrock ridges on the former bay floor, covered by modern lagoon salt deposits. Submerged outcrops of Cretaceous rock in the region were expressed in isobaths on maps of the 1950s as sublatitudinally extended submarine positive relief forms.

Secstan Island, which in 1954 was still separated from the shore by water, found itself deep in the center of the dried salt plain zone. The former Soviet Bay, bounded by Jangy-Su Spit, was completely dried up. From the distal end of this spit, the shoreline retreated to the northwest by about 10 km. In the southwest corner of the Gulf of Kara-Bogaz-Gol three peninsulas appeared, similar in structure to the Omchali Peninsula. They formed where local outcrops of Baku (lower Pleistocene) and possibly older (Upper Khazar) shell limestones were attached to the shore. At the end of the 1950s these remnants of eroded ancient beach ridges had been in the recently dried strip of wet salt flats. The shallow bay bounded by the small Ayman-Tubek Spit, which complicated the eastern edge of the South Kara-Bogaz barrier, disappeared entirely. After the bay dried up and the spit was attached to the shore along its entire length, the width of the proximal part of the South Kara-Bogaz barrier more than doubled.

A 1970 space photo (Figure 16.3) also shows some consequences of the Caspian level drop on the sea side of the Kara-Bogaz Spits: in places (particularly north of the inlet entry) ridges of Upper Khazar limestones, which previously had been submerged, projected from the water. In a 1976 space photo (not shown here) the contours of the open gulf waters remained roughly the same, even duplicated in certain details. However, the bay-floor drying continued and led to a marked expansion of Omchali Peninsula, for example, and to smoothing of the shoreline east of it. By 1975 the inflow of seawaters into Kara-Bogaz-Gol had decreased to 5.0 km^3, and the total area of its water surface had decreased to 8200 km^2.

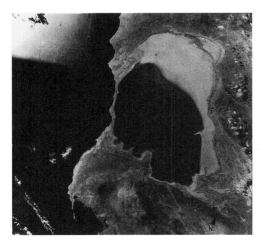

Figure 16.3 Photograph of the Gulf of Kara-Bogaz-Gol taken from Soyuz 9 spacecraft, 1970.

The regression of the Caspian Sea reversed in 1978. Sea level began to rise gradually, leading to increased inflow of Caspian water into Kara-Bogaz-Gol. By 1980 the water level in the gulf was 3.5 m below Caspian sea level (−28.448 m), and the gulf water area was 9500 km² (Bortnik and Luchkov, 1988). However, to save Caspian water and prevent excess evaporation, it was decided to block Kara-Bogaz Inlet with a nonoverflow dam. This dam was erected at a distance of about 0.5 km from the sea entrance into the inlet in March 1980. After the gulf was isolated, rapid degradation ensued. A series of space images taken from the Soviet Meteor 30 satellite in 1981, 1983, and 1984 recorded the reduction in gulf water area from roughly 3600 km² in 1981 to complete desiccation of the gulf and its transformation into a dry salt flat in 1984 (1984 photo shown in Figure 16.4).

The separation of the gulf from the sea did little to raise the Caspian level but did lead to a sharp change in physical and chemical conditions for formation of industrially exploited brine from the gulf. This necessitated a renewal of the Caspian inflow into the gulf. In September 1984 a controllable water inlet structure was built by laying 11 large-diameter pipes through the dam structure, designed to pass as much as 2 km³/yr. In the first years about 1.5 to 1.6 km³/yr of water was fed into the gulf.

During the 4.5 years of complete separation of the gulf from the Caspian, there was significant smoothing of the bottom from salt deposition. The bottom rose by 1 to 1.5 m in places (Bortnik and Luchkov, 1988). Seawater newly entering the gulf flowed freely over the smooth surface of the salt layer and in 1984–1987 caused great variability in gulf water area, especially under the influence of wind-induced surge phenomena. Sharp seasonal fluctuations of water volume played a significant role in areal fluctuation—maximal amounts were reached by the start of the evaporation season (May) and minimal at the end (October) in connection with desiccation of the shallowest southeastern part of the gulf. Thus, in March 1986, the water surface area of the gulf was around 4000 km², whereas in September of the same year it was only 1200 km².

A 1991 space photo from the Meteor satellite (scale 1:3,000,000, not shown) recorded a significant amount of water in the Kara-Bogaz-Gol basin, with an area of a little less than 2800 km², abutting the Kara-Bogaz barrier with its inlet. In shape,

Figure 16.4 Photograph of the Caspian from the Meteor 30 satellite, 20 August, 1984, showing complete desiccation of the Gulf of Kara-Bogaz-Gol (arrows).

the reservoir roughly mirrored the outlines of the whole Kara-Bogaz-Gol basin; it had a large arm east of Omchali Peninsula and very uneven shoreline in the east, which was probably associated with the presence of wind-surge troughs there.

Space photo STS045-81-57 (Figure 16.5), taken in March 1992, showed that the gulf surface was divided into three large parts. The eastern peripheral portion, approximately to the middle of the distal end of Jangy-Su Spit, is a dry salt plain (1) bounded on the gulf side by a salt scarp whose outlines show the salt cape described above (in 1970s photos). The western part of the gulf, directly abutting the Kara-Bogaz Spits, is distinguished by its deeper blue color (3) and evidently corresponds to the deeper parts of the gulf, filled with water entering through the pipes in the dam. Between these two parts a band of wet salt is visible, covered by a thin layer of brine (2). Overall, the situation in the gulf in 1992 is reminiscent of the situation in the 1970s.

However, the restoration of the physical and chemical properties of the gulf proceeded slowly, and in 1992 at the instruction of the government of Turkmenistan, dam destruction was begun. The area covered by gulf water had already expanded markedly by 1993. This is evident in space photos (scale 1:1,300,000, not shown) taken from a Russian Kosmos satellite in different spectral bands. The photos cover the western half of the Kara-Bogaz-Gol basin. In the south of the basin, water already washes the shores of the four peninsulas described earlier (Omchali and three others). In the north of the basin, water enters the bay bounded by the Karasukhutskaya Spit. The near-infrared band shows extensive shallows in the northern part of the gulf, and the infrared band shows small sections of shoals around all the southern peninsulas.

16.4 PRESENT-DAY KARA-BOGAZ-GOL

A series of color space photos taken from 1994 onward from the Space Shuttle and *Mir* illustrate the restoration of Kara-Bogaz-Gol had to its dimensions of the 1920s and 1930s (e.g., NASA photo STS059-L17-73, Figure 16.6). In the north and east the shoreline has moved up to the old inactive shore cliffs. The bay bounded by the Karasukhutskaya Spit in the northwest and Soviet Bay in the southeast has nearly

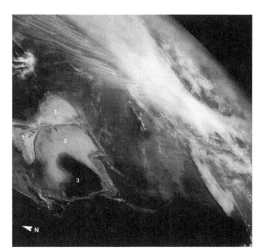

Figure 16.5 NASA photograph (STS045-81-57) of the Gulf of Kara-Bogaz-Gol and the eastern coast of the Caspian Sea, taken March 25, 1992. Numbers keyed to explanations in text.

filled with water. The shallowest southeastern corner of Kara-Bogaz-Gol is covered with water. Secstan Island has again acquired island status and is separated by a narrow inlet from the shore. Taraba Island is still connected to the shore, however. Two small islands appeared on the extension of the Karasukhutskaya Spit.

Within the bounds of the gulf (see Figure 16.6) one can trace the western and deepest part (1) abutting the Kara-Bogaz Spits from its deeper blue color, where dissolution of recently deposited salts is evidently taking place. In the shallower eastern part of the gulf (2), through the water layer one can see several extended salt scarps (3) with uneven outlines, parallel to the eastern shore (3). They probably record successive stages of gulf desiccation. The sharp salt projection south-southwest of Cape Kulan-Gurlan noted in the 1970s space photos is now completely covered by water again and can only be weakly traced in some photos (4). Along the northern shoreline on photos, one can discern a bright narrow fringe possibly corresponding to the removal of terrigenous material as a result of renewed shore erosion (5). On the surface of the salt bottom in the limits of the southern and southeastern shallows one can see a series of erosive chutes and troughs (6), in all likelihood associated with wind-induced surge currents.

Cape Kulan-Gurlan, which projects noticeably past the general line of the coast and is at the mouth of a large branching gully system, very visible in the photo, is the most outstanding element of the eastern shore. South of the cape, along the coast, there extends a bright strip of young shore barrier, behind which in places one can

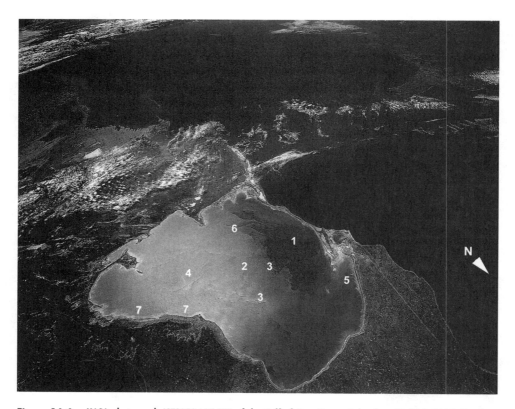

Figure 16.6 NASA photograph (STS059-L17-73) of the Gulf of Kara-Bogaz-Gol, taken April 14, 1995. Numbers keyed to explanations in text.

spot a narrow lagoon (7). The appearance of this complex—the large shore barrier with lagoon behind it—is generally very typical of depositional coasts of the transgressing Caspian Sea (Ignatov et al., 1993).

A similar topographic complex (1) on the sea side of the Kara-Bogaz Spits is shown in a more detailed picture in space photo STS068-247-5 (Figure 16.7). During regression such topographic forms were not seen here. In this same photo, a delta is visible near the interior mouth of the inlet (2). This relief form had also been observed earlier at other water levels in the gulf. The sand, shells, algae, and living organisms swept through the inlet by the current are deposited in the form of shoals and spits which together form the delta. According to Leontyev (1961), by 1957 the perimeter of the delta was 10 km; at present it is no more than 7 to 8 km and extends roughly 3 km into the gulf. On the sea side of the Kara-Bogaz Spits, from the lighter color one can trace a submerged ridge of Khazar (middle Pleistocene) shell limestone, which comes to the surface at the distal end of the northern spit to form a bedrock cape. Farther south its submerged continuation is also visible (3).

The space photos of the Russian-U.S. series also provide excellent images of specific geomorphic features of the gulf coast. From the different shades of color on photos one can map the surface of the Kenderly-Kayasan Plateau in the north and the staircase of Pleistocene wavecut and sedimentary marine terraces of different ages in the southern (and some in the eastern) edges of the gulf. The very smooth outline of the southern shore of Kara-Bogaz-Gol is also conspicuous, perhaps influenced by tectonic fractures, in addition to erosion.

16.5 CONCLUSION

The Gulf of Kara-Bogaz-Gol has a long and complex history, closely associated with the behavior of Caspian Sea level. In the last century the gulf has experienced three stages in its evolution caused by both natural factors (fluctuations in Caspian level) and human factors (artificial separation): a stage of significant water filling (1900–1980), although dropping sea level, reduced the inflow drastically from 1929 to 1980; a stage of complete desiccation (1980 to 1984); and a stage of renewed fill-

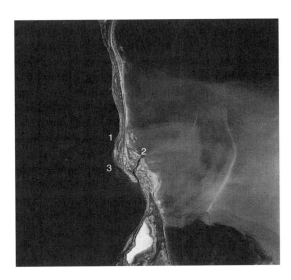

Figure 16.7 General view of the Kara-Bogaz Spits and inlet, October 9, 1994. (NASA photograph STS068-247-5). The delta at the inlet mouth and the turbid debris cones from it are visible. Numbers keyed to explanations in text.

ing (since 1985). The geomorphological consequences of the hydrological changes in the gulf are recorded in space photos from various years and of various types. After restoration of the link to the sea (partially in 1984 and completely in 1992), the gulf began to be quickly replenished with seawater. This process was greatly promoted by the present rise in the Caspian Sea level (by nearly 2 m since 1978).

The present state of the gulf and its shores is well reflected in a series of color space photos taken from the Space Shuttle and *Mir*. The reaction of Kara-Bogaz-Gol shorelines (as of all the Caspian Sea) to the rise in water level may be of interest in light of the expected increased rise in global sea level in the twenty-first century.

Restoration of the Gulf of Kara-Bogaz-Gol offers real opportunities for the rebirth of a unique natural landscape and ecological complex of the region, and preservation and use of unique natural brine resources. The significant volumes of seawater inflow into the gulf in 1992–1995, reaching a total of 132 km^3, have helped slow the rates in the modern rise in the Caspian Sea level by more than 30 cm; this has helped to reduce the urgency of the flooding problem. Space photos will allow us to record the history of the Gulf of Kara-Bogaz-Gol and to ascertain the details of the changes.

ACKNOWLEDGMENTS

We thank V. N. Bortnick and F. S. Terziyev from the Oceanographic Institute of the Russian Academy of Sciences for their valuable advice during preparation of this paper. We appreciate the efforts of TechTrans International, Inc. (Houston) in translating the manuscript into English. L. Prejean, C. A. Evans, and J. A. Robinson assisted in preparing illustrations. We also acknowledge M. A. Helfert and staff members of the NASA Office of Earth Sciences for constructive reviews of the paper.

REFERENCES

Bortnik, V. N., and Luchkov, V. P. 1988. Changes in the condition of the Gulf of Kara-Bogaz-Gol under conditions of restricted sea-water nourishment [original in Russian]. *Meteorologiya i Gidrologiya*, 9:113–119.

Dzens-Litovskiy, A. I. 1959. Geology and hydrological conditions of new types of sulfates in the Gulf of Kara-Bogaz-Gol [original in Russian]. *Works of the Oceanographic Commission, USSR Academy of Sciences*, V:305–313.

Ignatov, E. I., Kaplin, P. A., Lukyanova, S. A., Solovieva, G. D. 1993. Evolution of the Caspian Sea coasts under conditions of sea-level rise: model for coastal change under increasing "greenhouse effect." *Journal of Coastal Research*, 1:104–111.

Leontyev, O. K. 1961. History of Kara-Bogaz-Gol Gulf coast formation [original in Russian]. *Works of the Oceanology Institute, USSR Academy of Sciences*, XLVIII:34–66.

Mironova, N. Ya. 1959. Inflow into the Gulf of Kara-Bogaz-Gol and the change in the hydrological regime of inlet and gulf [original in Russian]. In *Problemy Kaspiyskogo morya (Problems of the Caspian Sea). Works of the Oceanographic Commission, USSR Academy of Sciences*, V:146–150.

Terziyev, F. S., Goptarev, N. P., Bortnik, and V. N. 1986. The problem of the Bay of Kara-Bogaz-Gol [original in Russian]. *Vodnyye Resursy (Water Resources)*, 2:64–71.

Chapter

17

Eddy Formation in the Caspian Sea

L. M. Shipilova
Department of Geography
Moscow State University
Moscow, Russia

ABSTRACT

Eddy movements of water masses along the shores of the Caspian Sea were discovered in photographs taken from U.S. and Russian spacecraft. Possible causes for the onset of rotary motions of the waters were analyzed. The most important conditions for stability disruption of the main littoral flow are regional surface wind flow, bathymetry, and specific features of the coastal topography. Other destabilizing factors include internal waves and littoral upwelling, which decreases the vertical stability of the nearshore waters by a factor of 7 to 8 in comparison with the open sea.

17.1 OBSERVATIONS OF EDDY FORMATION

Satellite observation is one of many tools for studying the marine environment. The most commonly used satellite data for oceanographic studies are in the optical and infrared electromagnetic wavebands. Images from space have recorded rotary (eddy) motions of water masses in many regions of the world's oceans. These include individual eddies and series of cyclonic and anticyclonic eddies varying size, both close to and far from the shore. Eddies are frequently identified in space photographs from the accumulations of phytoplankton and the spatial distribution of suspended sediments or finely ground ice. Infrared images show eddy structures from the temperature differences on the water surface. Not much is known about eddy formation in the Caspian Sea. However, sometimes individual eddies or series of eddies are readily visible on satellite and astronaut photographs.

17.1.1 Eddies Formed by Currents

Figure 17.1, taken with a hand-held camera from the Space Shuttle in November–December 1983, shows a large eddy near the southwest coast of the Caspian Sea, marked by debris from the Safid River, near Bandar-e Anzali, Iran (see location map, Figure 9.1). The bands of suspended deposits from the river mouth extend in the direction of the adjacent cape, indicating the flow direction of the littoral current. The first image shows high turbidity at the mouth of the river, indicating a high concentration of suspended sediment that decreases with distance from the river mouth. The width of the band of suspended sediment that borders the coast is uneven. In the section from the mouth of the main channel of the Safid River to the cape, the width of the turbid water is no more than 3 to 4 km, while south of the cape the width of the turbid zone increases roughly by a factor of 3, and some of the suspended material is drawn into the rotary motion of the waters, tracing an eddy with a diameter of 28 to 30 km.

The submarine slope in this region has the form of a steep step. Directly off the cape from the water's edge to a depth of 20 m is a gently sloping area 2.5 to 3 km wide consisting of fine and medium sands with some shell. The bottom slopes there are 0.004 to 0.005. Then the depths increase rapidly, reaching 450 to 500 m at a distance of just 10 to 12 km from shore. The steepness of the submarine slope increases by an order of magnitude.

Evidently, the speed of the nearshore current differs substantially from that of the offshore current at the shelf edge. The steep step of the submarine slope slows the movement of the flow near the shore, and the leading motion of the offshore current forces a turn toward the shore, leading to the formation of an anticyclonic eddy.

17.1.2 Eddies Formed by Surface Winds and Internal Waves

Eddy motions of water masses are encountered in other regions of the Caspian Sea. Figures 17.2 and 17.3 show a section of the western coast of the Caspian Sea in black-and-white space photos taken August 30 and June 30, 1979, respectively, from a Russian Kosmos satellite using a KATE-200 camera. The area covered in Figure 17.2 is a section of coast between Cape Satun (A, Figure 17.2) in the south, the Sulak River delta (B, Figure 17.2), and Agrakhan Peninsula (C, Figure 17.2) and Chechen Island (D, Figure 17.2) in the north. Figure 17.3 covers only the southern part of the Agrakhan Peninsula and the Sulak River delta. Both photos show littoral zone movements of suspended sediments in the form of streams parallel to the shore, and in the form of eddies.

Short-period wind-driven waves influence the movement of sediment only in the comparatively narrow littoral band and directly at the water's edge. But the deciding role in the transport of the sediment belongs to gravitational waves and eddy-current formations. The speeds of eddy movements over an area are not great, but the speeds inside the eddy are quite sufficient to stir up the bottom deposits and keep them in a suspended state for a long time, especially the fine sediments. The suspended material allows the eddies to be visible on photographs.

The patterns of suspended sediment in the littoral zone on June 30 (Figure 17.3) and August 30 (Figure 17.2) differed substantially. These patterns are attributed, in

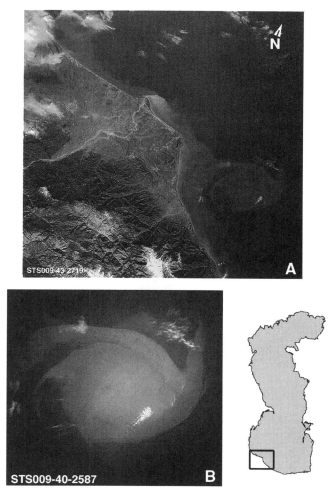

Figure 17.1 Southern Caspian coast in Iran: NASA photographs (A) STS009-43-2719; (B) STS009-40-2587. Sediment from the Safid River becomes entrained in an eddy off Cape Sefid-Rud.

part, to the prevailing wind and atmospheric pressure patterns at the times the pictures were taken. On August 30 (Figure 17.2), several eddies are visible along the coastal zone. A cyclonic eddy in the region of Makhachkala, with a diameter of around 10 km, is most clearly shown (1, Figure 17.2), along with a smaller eddy induced by it, located to the south, where the shoreline changes (2, Figure 17.2). Farther north, along the seaward side of the Agrakhan Peninsula, one can make out two more diffuse eddies. The first, more distinct eddy is directly in the region of the Terek "pioneer" delta (4, Figure 17.2), and the second eddy is between the Sulak and Terek River deltas (3, Figure 17.2).

The differences in the distribution of the suspended sediment seen in the June and August 1979 images might be explained by analyzing the maps of the ground and

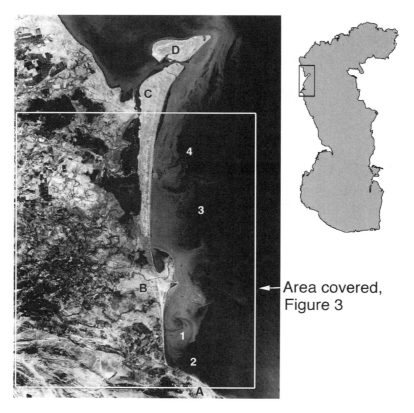

Figure 17.2 Movement of suspended particles offshore of Dagestan on August 30, 1979, photographed from Kosmos satellite. Eddies form in the absence of strong zonal winds. Geographical features mentioned in the text: A, Cape Satun, near Kaspiysky; B, Sulak River delta; C, Agrakhan Peninsula; D, Chechen Island. The Terek River is inland from the Agrakhan Peninsula. Numbered features discussed in the text: 1, cyclonic eddy in region of Makhachkala; 2, smaller eddy induced by it; 3, 4, diffuse eddies offshore from the Agrakhan Peninsula. The bright parallel bands along the southeast edge of the photo (lower right) are internal waves approaching Cape Satun.

Figure 17.3 Movement of suspended particles in the littoral zone off Makhachkala, June 30, 1979, photographed from Kosmos satellite. No eddies are observed. Zonal winds predominate.

high-altitude baric topography. However, in lieu of these materials we used another criterion to evaluate the type of atmospheric circulation: the zonal air-exchange index (I_z), a measure of the intensity of zonal (east-west) winds, proposed by Dmitriev et al. (1989). Figure 17.4 is a time-series plot of the zonal index (I_z) in June and August 1979. The I_z is plotted relative to data (historical climatological norms) from the elementary synoptical periods (ESP) calendar for 1949 to 1989. High positive values of the I_z anomalies are associated with intensive zonal wind fields and western airflow, while negative values of I_z are indicative of weakened zonal wind fields. In addition, a negative I_z slope from positive values indicates a transition from a cyclonic field to an anticyclonic field. We know that intensive zonal (east-west) airflow in the northern hemisphere is associated with cyclonic sequences, and interlatitudinal air exchange leads to the development of anticyclonic sequences (Dmitriev et al., 1989).

Figure 17.4 shows that the June and August zonal index dependence curves are out of phase. Whereas in August nearly all the values of the zonal index anomalies are below the zero line (norm), in June the zonal index anomaly dependence curve lies mostly above the zero line, and the positive values of I_z reach rather high values. This means that in June zonal (east-west) airflow predominated, while nonzonal airflows with reduced wind speeds were characteristic of August, improving conditions for eddy formation.

Although the surface winds are the primary factor for nearshore currents, eddy formation and sediment transfer, two aspects of the local nearshore bathymetry, are also important. First, in the section from Cape Satun to the Sulak River delta, the orientation of the shoreline changes from northwesterly to northerly (Figure 17.2). Second, in the littoral zone the bottom slopes increase by an order of magnitude in comparison with the deepwater slopes. I suggest that internal waves approaching along the normal to Cape Satun and shown in the photo (Figure 17.2) in the form of rectilinear and parallel bright bands also served as an impetus to eddy formation. The fact that there are no other visible signs of waves on the sea surface can also be considered as an indirect confirmation that this satellite photo does not show swells,

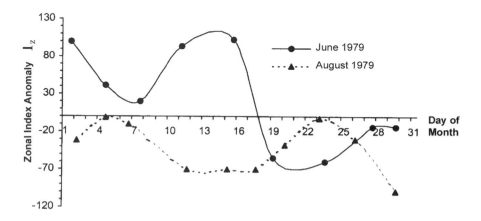

Figure 17.4 Time sweep of zonal index (I_z) anomalies in June (solid line) and August (dashed line) 1979. I_z is a measure of zonal (east-west) winds proposed by Dmitriev et al. (1989).

let alone wind-driven waves. In addition, the distance between the individual bright bands is measured in kilometers, while the average swell wave period in this region of the sea is no more than 10 to 12 seconds, and the maximal swell wavelength will accordingly be 160 to 200 m.

Internal waves arising in the discontinuity layer are rather frequent phenomena in the Caspian Sea. Ivanov and Konyaev (1976) recorded internal waves with a span of 2 to 4 m along this coast in conditions similar to those we observed. They used six horizontally separated sensors in the Caspian littoral zone at depths of from 24 to 34 m. Internal waves arose during a rapid rise of the discontinuity layer and subsequent slow drop, and the phenomenon was conditionally named a *thermocline bore*. The average bottom slope in the observation region was 0.001. Observation of internal waves lasted more than a week.

Analysis of the image in Figure 17.2 and the data from Ivanov and Konyaev (1976) have allowed us to track the transformation of internal waves off Cape Satun. Comparison of the distance between individual slicks with the bathymetric map showed that at depths of 30, 25, and 20 m, the lengths of the internal waves were 2.4, 1.6, and 0.6 km, respectively. I also calculated internal wave periods and propagation speeds for the three control depths (Table 17.1). As depth decreases, wavelength decreases, the period shortens, and most important, the wave propagation speed drops. When the wave energy transfer rate drops, conditions are created for resonance transfer into other types of motion (Brekhovskikh, 1975), particularly into eddy formation. Thus, one of the principal reasons for eddy formation in this part of the Caspian Sea is the configuration of the shoreline.

17.1.3 Eddies Formed by Upwelling Water

The phenomenon of littoral upwelling, the rise of cold water to the surface along the coastal zone, is common in oceans around the world. Until recently the structure and dynamics of littoral upwelling waters in the Caspian Sea had been studied incompletely. However, Arkhipkin (1990) found that sharp water temperature decreases were basically confined to the shores of the western and eastern central Caspian, and the frequency and duration of upwelling off the east coast was higher than along the west coast by a factor of 2.

A number of problems associated with rational use of natural resources along the Caspian coast require study of upwelling processes. As there are insufficient direct instrumental measurements of upwelling processes, important information can be

TABLE 17.1 Characteristics of Internal Waves, Cape Satun, Dagestan[a]

Depth (m)	Wavelength (m)	Period (s)	Propagation Speed (m/s)
30	2400	498	4.8
25	1600	336	4.8
20	600	222	2.7

[a]Lengths of internal waves were determined from Figure 17.2 and Ivanov and Konyaev (1976). Internal wave periods and propagation speeds are calculated.

obtained using the data from space images of Caspian Sea waters. Observations made in the infrared bands from AVHRR sensors on NOAA satellites are particularly useful (Figures 17.5 and 17.6; see the color insert).

Figure 17.5 shows the temperature distribution on the sea surface off the west coast of Mangyshlak Peninsula and south of the peninsula, obtained from the NOAA 11 satellite on October 4, 1991. A band of warm water occurs along the coast; in places the width of the warm-water band reaches 20 km. The image quality does not allow detailed examination of the temperature distribution pattern over all the littoral waters, but one can clearly see how individual streams of warm water mark the cyclonic eddy off the west coast of Mangyshlak Peninsula. South of the peninsula, where the shore contour has the form of a weakly concave arc, fingers of cold waters impinge on the shore, and warm-water plumes flow seaward, also creating a pattern of cyclonic eddies.

Figure 17.6 is a water-surface-temperature distribution map, in this case showing littoral upwelling off the eastern shore of the central Caspian on July 31, 1995. The map was compiled from AVHRR data on the basis of algorithms proposed in Planet (1979). The water surface temperature was estimated from the temperature differences at wavelengths of 11 and 12 µm. The temperature difference between the warmer offshore and colder nearshore surface waters is 10°C. The width of the cold-water band is uneven, which may be explained by shoreline irregularity (capes, coastal arcs) and the complexity of the bottom relief. In Figure 17.6 it is clear that the width of the cold-water zone is maximal in shore concavities and minimal at capes.

Conditions for upwelling include the appropriate atmospheric (wind) conditions and hydrological processes. Studies of recent decades have shown that upwelling is closely associated with entrained waves (Efimov et al., 1985). Entrained continental waves arising in the cape region may be the reason for the formation of upwelling centers seen in Figure 17.6. However, this still requires experimental confirmation.

Littoral upwelling decreases the vertical stabilization of upper-layer waters near the shore in comparison with the open sea by a factor of 7 or 8. The sharpest changes are observed 5 to 20 km from the shore (Arkhipkin, 1990). Together with the influence of shore and bottom topography, this leads to the formation of the eddy structures depicted on the map. Thus, in Figure 17.6, in the region of Yeraliyev Peninsula, one

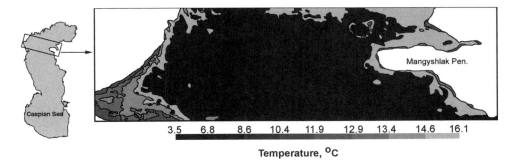

Figure 17.5 Surface water temperatures (°C) off of the Mangyshlak Peninsula on October 4, 1991, based on AVHRR data from the NOAA 11 satellite. Warm water bathes the coast; no upwelling is observed.

can quite clearly trace the cyclonic eddy structure, with diameter equal to roughly 23 to 25 km. A similar eddy but of somewhat smaller diameter (15 to 16 km) occurs off the southern coast of Mangyshlak Peninsula.

17.2 CONCLUSION

Ocean eddies are a common phenomenon in the Caspian Sea and are the result of the combined effects of different atmospheric conditions, hydrological parameters, and coastal influences. Here I have examined several instances of eddy formation in the Caspian Sea using a variety of observation tools in space: hand-held cameras, remotely controlled cameras, and radiometric scanners. Eddy movements affect the local currents, sediment dispersal, nutrient abundance, plankton, and pollution movements. Therefore, the study of eddy formation and eddy motion is extremely important for understanding the complex interactions between the land, the ocean, and the atmosphere in the littoral zone.

ACKNOWLEDGMENTS

The author acknowledges the efforts of TechTrans International, Inc. (Houston) in translating this manuscript into English, L. Prejean and C. A. Evans for assistance with illustrations, and M. A. Helfert and staff members of the NASA Office of Earth Sciences for constructive reviews of the manuscript.

REFERENCES

Arkhipkin, V. S. 1990. (Specific features of structure and dynamics of littoral upwelling in the Caspian Sea). In Kosarev, A. N., ed., *Kaspiæiskoe more: struktura i dinamika (Caspian Sea: Structure and Dynamics)*. Moscow: Nauka, pp. 61–74.

Borisov, E. V., ed. 1977. *Issledovanie turbulentosti I reshenie zadach perenosa zagriazniaiushchikh veshchestv v more (An Investigation of Turbulence and Resolution of Pollutant Transfer Problems in the Sea)*. Trudy Gosudarstvennogo Okeanograficheskogo Instituta (GOEN Works), No. 141. Moscow: Moskovskoe otd-nie Gidrometeoizdata, 171 pp.

Brekhovskikh, L. M. 1975. *Volny v sloistykh sredakh (Waves in Layered Media)*. Moscow: USSR Academy of Sciences Publishing, 42 pp. [Editor's note: Citation for English translation: Brekhovskikh, L. M. 1980. *Waves in Layered Media*, 2nd ed., Beyer, R. T., translator. New York: Academic Press, 503 pp.)

Dmitriev, A. A., Seltzer, P. A., Kontratyuk, S. I., and Kuchin, V. A. 1989. *Makromasshtabnye atmosfernye protsessy i srednesrochnye prognozy pogody v Arktike (Macro-scale Atmospheric Processes and Medium-Term Weather Forecasts in the Arctic)*. Leningrad: Gidrometeoizdat, 255 pp.

Efimov, V. V., Kulikov, E. A., Rabinovich, A. B., and Fayn. I. V. 1985. *Volny v pogranichnykh oblastiakh okeana (Waves in Ocean Boundary Regions)*. Leningrad: Gidrometeoizdat, 280 pp.

Ivanov, V. A., and Konyaev, K. V. 1976. Bor na termokline (Thermocline bore). ФАО [*FAO* (Food and Agriculture Organization, United Nations)], 12(4):416–423.

Planet, W. G., ed. 1979. *Data Extraction and Calibration of TIROS-N/NOAA Radiometers*. NOAA Technical Memorandum NESS 107-Rev.-1. Washington, DC: National Oceanic and Atmospheric Administration, 58 pp.

Chapter

18

Geomorphology of Southern Azerbaijan and Coastal Responses to Caspian Transgression

E. I. Ignatov and G. D. Solovieva

Department of Geography
Moscow State University
Moscow, Russia

ABSTRACT

We examined characteristic features in the development of the southwestern coast of the Caspian Sea. Space photographs were interpolated to construct a geomorphological map of this region, reflecting its tectonic and geomorphological differentiation. Through comparative analysis of a time series of space photographs, we distinguished coastal changes that have resulted from the rise in Caspian sea level over the last two decades.

18.1 INTRODUCTION

The study and comparative analysis of space photographs of the Caspian basin taken at different times since 1980, including relatively low and high sea levels, make it possible to detect the consequences of rearrangement of shorelines in conjunction with rising sea level. The present transgression of the Caspian allowed us to observe its influence on the development of various coastal types and track the response by the coastal zone. The practical and theoretical importance of studies of coastal transformation is evident, as stability of coastal land and beaches ensures economic feasibility and safety of their commercial use.

Our objectives were (1) to describe the geomorphology of the coastal region around the Kura River outflow, including seafloor topography; and (2) to investigate dynamic coastal responses to rising and falling sea levels in this region. We interpret and compare photographs taken from 1980–1996 from a variety of space platforms: the Meteor 30 satellite, Space Shuttle, Kosmos satellite, and *Mir* space station. We first combine the photos to gain information about underlying geomorphology and then use the photos as a time series documenting coastal change associated with Caspian sea-level rise.

18.2 GEOLOGICAL AND GEOMORPHOLOGICAL CHARACTERISTICS

Our study area is the southwestern coast of the Caspian Sea, in particular the low mountain ranges and foothills of the extreme southeasterly spurs of the Greater Caucasus Mountains, the eastern Kura depression and Talish Mountains bordering it on the south, and the Lenkoran depression (Figure 18.1). We identified structural features of the subaerial and seafloor topography on the basis of space photographs and constructed a geomorphological map of southeastern Azerbaijan (Figure 18.2; see the color insert). We also include a high oblique photograph taken from the *Mir* to provide regional context (Figure 18.3). Here we present detailed interpretation and comparative analysis of multitemporal space photographs from 1980 (Figure 18.4), 1990 (Figure 18.5), 1993 (Figure 18.6), and 1996 (Figure 18.7) from the Meteor 30 satellite, Space Shuttle, Kosmos satellite, and *Mir* space station.

As a whole, this segment of the Caspian coast is characterized by rectilinear, tectonically controlled topography in which modern features have been created by neo-

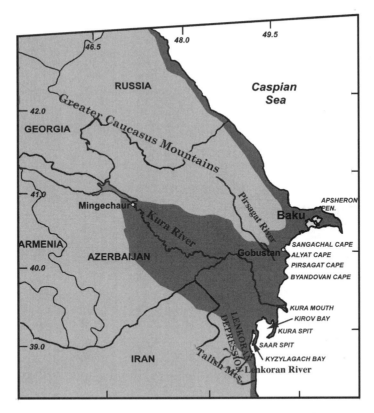

Figure 18.1 Location map of the coastal study area and surrounding region, emphasizing areas mentioned in the text.

Figure 18.3 Synoptic photograph of Azerbaijan taken from the *Mir* space station, June 1996 (NASA photograph NM21-757-053). The Apsheron Peninsula is the feature in the center of the photo. Southward down the coast, a lighter region of hills can be seen and then the darker lowlands of the Kura depression.

Figure 18.4 Photograph of the southwestern Caspian coast from Pirsagat Cape to the Kura River delta, Kura and Saar Spits from the Meteor 30 satellite, October 1980. This photograph represents a baseline before the effects of Caspian Sea transgression began to be observed.

tectonic movements and modified by other processes (fluvial, aeolian, solonchak, and deflationary, but primarily marine).

18.2.1 Kura Depression

A large central part of the coastal territory consists of the Kura depression, which corresponds to the tectonic depression of the same name, a zone of maximum neotectonic downwarping beginning in the upper Pliocene (Akchagyl period). The flat floor of the Kura River valley and coastal plain has formed by active accumulation of unconsolidated material, derived from both terrestrial and marine sources (areas 11–16, Figure 18.2). The map reveals coastal forms left behind by the sea in historical times: coastal barriers, cliffs, and terraces (areas 30 and 31, Figure 18.2). Segments of former lagoons have been modified by solonchak–deflationary and aeolian

Figure 18.5 Photograph of the same segment of coastline shown in Figure 18.3, taken from the Space Shuttle, March 1990 (NASA photograph STS036-89-051).

Figure 18.6 Photograph of the same segment of coastline shown in Figure 18.3 and 18.4 taken from the Kosmos, 1993. These three photos encompass the most dramatic transgression effects on the Kura River Mouth and Saar Spit.

processes (area 12, Figure 18.2). The map shows ancient river valleys, such as the Kura (areas 13 and 28), thick detrital cones (areas 9 and 26), and ancient deltas (areas 14 and 28). According to drilling data, the Quaternary deposits reach a maximum thickness of nearly 1400 m in the Kura Spit region and 700 m on the Saar Spit (locations in Figure 18.1).

Along the margins of the Kura Spit in the depression, Quaternary marine transgressions have produced a staircase of 14 wavecut sedimentary terraces (Shirinova, 1975), not counting fragmentary sections. The photographs reveal that the Kura depression comprises a general plain interrupted in the northern part by low, oval hills. These hills are peripheral southeast spurs of foothills that appear as elongate light spots on photographs. This is perhaps best viewed in a synoptic regional photo (Figure 18.3; see also Figure 18.7).

The central and lowest portion of the Kura depression is occupied by the pre-Kura alluvial–lacustrine depositional plain with an abundance of intricate, meandering bends, oxbows, and ridges along the riverbed (Figure 18.2). This lowland strip is 8 to 10 km wide and is characterized by an undulating pattern on the images (see Figures 18.4 and 18.6). Traces of the prior Kura riverbed are visible as a chain of large, swampy depressions elongated in a southeastern direction. The southern

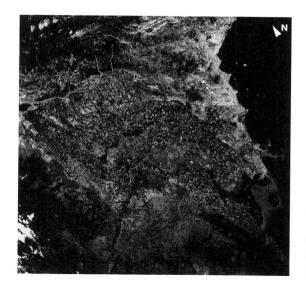

Figure 18.7 Photograph of the Gobustan coast, Kura depression, Kura River delta, and Kura and Saar Spits taken from the *Mir* space station, July 1996 (NASA photograph NM21-772-085). Erosional process continued relative to Figure 18.5. Vegetative responses to transgression are also apparent in the area of the Kirov and Kyzygilach bays.

part of the Kura depression is a very flat plain (seaward slopes no greater than 1 to 2°) composed of a sequence of deltaic and shallow-marine deposits. The series of old Kura deltas—of which there are at least three—was formed at the time the river flowed into the Kyzylagach Bay (Figure 18.2). At present, this region is undergoing intensive agricultural development; the ancient deltas are being tilled and their configuration is revealed by the orientation of the boundaries of agricultural lands (Figure 18.7).

18.2.2 Gobustan Foothills

The northern outline of the Kura depression, the Gobustan range of low mountains, consists of middle to upper Pliocene sandy, argillaceous, limey deposits. The territory is well exposed, so that finer details of the morphological structure are clearly visible on the photographs as the combination of segments with different photographic hues (Figure 18.7). The mountains are divided into denudation–erosion, arid–denudation, and mud–volcano types. Submontane strips border the foothills. Peculiarities in the geological and tectonic structure, along with the arid climate of the region, gave rise to characteristic topographical features, specifically the predominance of structurally denuded low mountain ranges and flat plains of slope-wash, alluvial-fan, and marginal-marine origin. The latter are readily visible as small capes along the shore (Figures 18.1 and 18.7). In general, morphological structures become less distinct and elevations diminish toward the Caspian Sea basin.

The Gobustan coastal segment consists of a system of terraces formed by both erosion and deposition; they appear on photographs as a solid, light-gray band nearly 10 km wide along the sea (Figure 18.7). Submontane strips of alluvium, alluvial-fan deposits, and slope wash line the foot of the mountains and produce tectonically controlled depressions. An example of such a consequent synclinal depression in the lower Pirsagat River valley (large alluvial fan shown in yellow in Figure 18.2) can be seen in the photographs as an extremely light area inland from the Pirsagat Cape (at the top of Figures 18.4, 18.5, and 18.7).

The low mountain ranges mentioned above represent a unique region of the world where mud volcanoes and genetically related topographic features have developed (area 25, Figure 18.2). Concentric light spots on the photographs are interpreted as volcanic cones. Mud volcanoes form due to rapid tectonic movements and the eruption of shattered rock and sediment with pressure from petroleum gas. Many of the volcanic cones are confined to a deep fault (the Adzhichai-Alyat) and its southeastern branches.

18.2.3 Talish Mountains

On the southern margin of the Kura depression the Talish Mountains appear on the photographs as a characteristic pattern of highly differentiated regions (Figure 18.7). The deeply incised erosional valleys and alluvial fan/slope wash accumulations of submontane plains are readily visible. The mountains encompass a piedmont depression to which the Lenkoran River valley is confined, along with deluvial–proluvial washout slopes ("badland" type; area 5, Figure 18.2).

18.3 MODERN COASTAL CHARACTERISTICS

The southern shoreline consists of three large sedimentary forms: the nose of the Kura River delta and the Kura and Saar spits to the south of it (Figure 18.1). A comparison of multitemporal space photographs reveals that significant changes in coastline configuration took place precisely in this segment during the last period (Figures 18.4 to 18.6). The 1980 photograph (Figure 18.4) depicts the coastline when the Caspian Sea was at rather low levels, reached after slow decline over a prolonged period. In 1977, the Caspian was at the lowest level of the century (−29.02 m). From that time levels began to rise continuously and, at first, rapidly. Levels have increased by more than 2 m in the last 20 years.

The change from regressive to transgressive conditions triggered significant changes in coastal-zone dynamics in the study area. Because sedimentary forms are the most sensitive indicators of changes in the coastal zone, we would expect that Caspian sea-level rise would influence the development of depositional facies. These changes are reflected on two other photographs taken when the transgression was already at its height (Figures 18.3 and 18.7). As a whole, one can note increased erosion not just on former erosional coasts but also of several depositional structures. This relates not only to rising sea level, but also to intensive human activity on the coasts and river valleys in the Caspian basin.

18.3.1 Kura River Delta

The present Kura River delta started to form in the early nineteenth century, when the river broke through the coastal barrier that existed along the outer margin of its shallow estuary and began building a new delta to the east of it. At the outset and until 1957, the delta advanced seaward, growing at a rate of 50 to 60 m/yr, corresponding to an annual increase in area of nearly 1.5 km². Kura accretions were evidently fed not just from the marine area of this river delta, but also from neighboring sedimentary portions of the coast, in particular from the Kura spit.

Construction of the Mingechaur hydroelectric station in the river valley in 1953 (Figure 18.1) and the associated decrease by almost half in sediment load carried by the Kura River (from 43 million tons to 20 to 25 million tons per year) caused intensive delta erosion starting in the early 1960s. This process was exacerbated by a stabilizing, rather than falling, sea level. According to data from Mekhtiev (1966), the eastern portion of the delta was being eroded at a rate of 15 to 20 m/yr during that time. However, the continuous (until 1977) drop in sea level resulted in progradation of the delta as it followed the receding coastline. Based on the Meteor photograph (Figure 18.4), the maximum length of the delta lobe reached 24 km (3 to 10 km wide) in the late 1970s and early 1980s.

By the 1990s, photographs show that the length of the delta had decreased to 12 to 13 km (Figure 18.6). Of course, due to the rise in sea level, flooding of low-lying coastal lands also played a large role in the modification and erosion of sedimentary structures. Transgressive changes are most evident in one of the older northward branches of the Kura that is currently submerged (Figure 18.7).

The movement of suspended material at the mouth of the Kura River also can be interpreted from the space photographs. Turbulent plumes of muddy water in the shallows point to a general southward tendency for coastal current movement (Figures 18.3 and 18.4).

18.3.2 Kura Spit

The Kura spit also experienced a complicated pattern of development in this century, having first acted as a coastal bar, then a peninsular sandbar fed by sediments from the Kura River (Figure 18.4). Today, it has been transformed into islands (Figure 18.7). As with the delta, sandbar erosion began when sediment discharge from the Kura River was eliminated after construction of the Mingechaur reservoir. The growing delta lobe and its expansion into greater depths of the open sea reduced shallow sediment transport to the sandbar and initiated its destruction during the regression. Destruction was most severe on the east side of the sandbar, which is exposed to prevailing northeasterly winds. The average rate of coastal erosion over a 10-year period before the transgression was 6 m/yr, reaching 25 m/yr in some years. The distal edge of the sandbar increased both in length and width due to sediment transport along the sandbar (Leontev et al., 1987).

The rise in sea level helped initiate the washing out of the Kura Spit, and coastal destruction by waves in the mid-1980s caused seawaters to break through to Kirov Bay and transform the southern end of the spit into a string of islands. Between the 1980 and 1993 (Figures 18.4 and 18.6), a 10-km segment of the Kura Spit was completely destroyed—a strait now exists in its place. The distal portion of the spit narrowed by almost half (from 6 or 7 km to 3 km). The width of the Saar Spit also narrowed significantly during this time. These changes appear to be associated with the flooding and submergence of coastal sections of low modern terraces and of the wide tidal zones which almost completely bordered the coasts of the Kirov and Kyzylagach bays during the regressive period.

The 1996 photograph (Figure 18.7) shows that degradation of the new depositional landforms (Kura and Saar Spits) continues, even though sea level has stabilized in recent years and transgression rates have decreased. In 1995 to 1996, a negligible drop in the sea level was noted in comparison to preceding years. However, these short-term changes in sea conditions have not yet influenced the course of coastal processes. Changes in erosional and depositional regimes are known to exhibit a certain inertial delay, as did the onset of morphological change from fluctuations in reservoir level.

The width of the strait separating the distal part of Kura Spit increased by another 2 to 3 km, while the width of the Saar spit narrowed greatly. A dark band extends along the coastline of the Kirov and Kyzylagach bays, evidently indicating flooded portions of the most recent terrace surfaces (1929 and 1940), as well as former tidal zones currently covered by reeds.

18.3.3 Gobustan Coast

The outline of the northern coastline (Gobustan area) comprises several arcuate sectors that are concave to the sea and separated by Sangachal, Alyat, Pirsagat, and Byandovan capes. These capes stem from anticlinal folds whose continuation offshore is characterized by complex bottom topography and the islands and shoals known as the Bakin Archipelago. The islands block the coast and generate an undulating shadow, thereby controlling significantly which sectors of the coast are formed by abrasion and which by accumulation. Tectonic dislocations extending along the axes of anticlinal zones from the coast to the sea are responsible for the high number of mud volcanoes, both on the capes and adjacent islands.

Transgressive changes in depositional features are also mirrored in the extensive development of lagoons, which have formed on low surfaces of young terraces and are separated by young sandbars transformed from coastal barriers. These segments are visible north of the Kura River delta and also between the erosional capes of the Gobustan area (Figure 18.7).

The capes themselves (mud volcanoes, as a rule) were bordered by erosional platforms (benches), whereas the cliffs were not reached by the waves. Erosional processes interrupted by regression have now been revived. The benches are now flooded by the sea, and storm waves reach the bases of the cliffs, especially during wind surges.

18.4 CONCLUSION

The processes of coastal transformation that we have observed along the southwestern Caspian coast are representative of changes in other regions of the Caspian coast, with several local variations. The elimination of sediment-bearing river runoff due to dam construction also resulted in erosion of deltaic coasts of the Volga, Ural, Sulak, and Terek Rivers. Sections of the shore that were fed by sediment from these rivers have been similarly eroded. Rising sea levels intensified this erosion throughout the Caspian.

Lagoon formation is another process typical of many depositional sectors of the Caspian seashore. Lagoons border the southeastern coast and other significant portions of the Caspian coast in Dagestan and Azerbaijan. Studying changes in the dynamics of Caspian coasts is important not just from a practical standpoint for human activity, but also from a theoretical standpoint. Such observations can be very significant for understanding the evolution of coastal zones of the world ocean and global tendencies in their development. In fact, the level of the world ocean is now rising, and correspondingly, coastal zones around the globe are in transgressive modes of transformation.

ACKNOWLEDGMENTS

The authors acknowledge the efforts of L. Prejean and J. A. Robinson in graphics preparation and TechTrans International, Inc (Houston) for translating this manuscript into English. M. A. Helfert and the staff members of the JSC Office of Earth Sciences provided reviews and editorial assistance.

REFERENCES

Leontev, O. K., Lukyanova, S. A., Solovieva, G. D., Veliev, Kh. A., and Ignatov, E. I. 1987. (Modern-day washout of accumulational coasts of the Caspian Sea). In Prirodnye osnovy beregozashchity (*Natural Principles of Coastal Protection*). Moscow: Nauka, pp. 91–99.

Mekhtiev, N. N. 1966. *Dinamika i morofologiya zapadnogo poberezh'ia uzhnogo Kaspiya (Dynamics and Morphology of the Western Coast of the Southern Caspian)*. Baku, Azerbaijan: Academy of Sciences of Azerbaijan SSR, 111 pp.

Shirinova, N. Sh. 1975. Noveishaia tektonika i razvitie rel'efa Kura-Araksinskoe depressii (*The Latest Tectonics and Development of the Kura-Araksin Depression*). Baku, Azerbaijan: ALM Publishing House, 188 pp.

Chapter
19

Land-Use Changes in the Northwest Caspian Coastal Area, 1978–1996: Case Study of the Republic of Kalmykia

A. S. Shestakov
Russian Academy of Sciences
Moscow, Russia

ABSTRACT

Loss of farming and range land along the Caspian coast of Kalmykia has been documented by means of space photographs, field observations, and statistical data. Between the onset of the current Caspian transgression in 1978 and 1996, cultivated land in the Lagan region had been diminished by 48% and grazing land by 76.8%. Marshes and submerged lands have increased by 460 and 700%, respectively. Some changes, such as increases in submerged and marshland and in brushy cover, are easily observed on conventional aerial and space photographs.

19.1 INTRODUCTION

Study of land use and dynamics is one application of astronaut–cosmonaut photographs. These data are especially useful for representing real-time changes and for monitoring inaccessible areas. As with any other remote-sensing method, it is best to use space photographs in combination with field observations and statistical data.

I studied land-use changes resulting from sea encroachment along the Caspian coast in the Lagan region of the Republic of Kalmykia (Figures 19.1 and 19.2; see the color insert for Figure 19.2). I conducted field reconnaissance to perform a visual survey of the flooded areas, to examine engineering structures erected to protect local communities and to collect locally available statistical data. Combining these data yielded a better insight into the significance of the space photographs obtained in the course of the Shuttle–*Mir* Earth observation project.

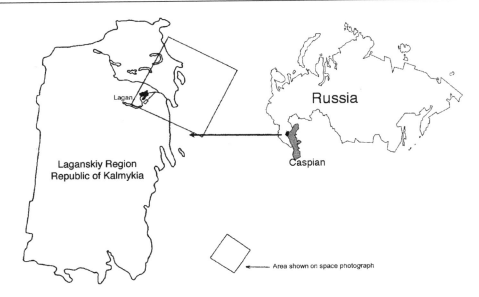

Figure 19.1 Location map of the study area: Lagan Region, Republic of Kalmykia.

Years of receding seas have permitted ongoing development in the Caspian coastal zone, including construction and utilization of newly emergent areas. That particular period of time coincided with the highest level of industrial, agricultural, and land development in the country. Communities, spas, and hotels were built along the coast, along with the supporting transportation and power infrastructure. Oil production operations began in a number of the coastal areas.

The rapid rise in sea level that started in 1978 immediately caused complex environmental, social, and economic problems. Along the northern coast of the Caspian the situation is aggravated by short-lived but frequent wind-tide flooding of hundreds of square kilometers of the coast. The vicinity of Lagan (Figure 19.1), also known as Kaspiyskiy, is one of the windiest areas of the northwestern coast of the Caspian; winds in excess of 14 m/s are recorded 40 days a year, and as many as 70 days in some years. Typical wave height at the coast of Kalmykia is between 0.9 and 1 m. Between June 1, 1993 and June 30, 1995 alone, sea level on the Lagan bank rose from −26.9 to −26.3 m, which is more than twice the average encroachment rate throughout the Caspian. The latest tidal surge on the Kalmyk sector of the coast was from March 4 to 7, 1995, when sea level in Lagan bank channel rose from 1.87 m to 2.75 m. That surge overwhelmed protective structures and flooded considerable areas of land, as well as several city blocks in Lagan and streets in the township of Jalykov. Thus, the 1995 surge was a strong factor influencing subsequent land-use change.

19.2 METHODS

This work was based on field studies, cartographic materials, aerial photographs, space photographs, and regional statistical data. Statistical data were provided by

(1) the State Committee on Land Resources of the Republic of Kalmykia, (2) the Republican Statistical Committee, (3) the Committee on Land Resources of the Lagan Region, and (4) the Lagan Region Environmental Protection Committee. Cartographic materials and aerial photos were provided by the Caspian Region subsidiary of the Priroda State Scientific Research and Production Center. Space photographs were provided by Johnson Space Center, NASA.

Photographs are especially important for surveying land use in this area because the coast has become swampy and inaccessible due to rising water. I documented land-use change, caused by rising Caspian Sea level on the Kalmykia coast, using color space photographs obtained in the course of several missions. (A regional view of the Volga delta is shown in Figure 19.2, a more detailed photo in Figure 19.3; see color insert for both.)

Space photographs were used for general clarification of the situation in the coastal area in comparison with statistical data and materials from aerial photographs. The data presented in Tables 19.1 and 19.2 and Figure 19.4 are based on the statistical information from the cited reference organizations and do not take into account the results of the photographic interpretation. The space photographs were checked in the field in order to work out a classification for subsequent decoding. Archived materials, maps, and statistical data available from local authorities (see the list above) supplemented remote-sensing and visual surveys. Research results were further verified using aerial photographs (scale 1:50,000) of the region. Complete integration of information from field reconnaissance and from interpretation of space photographs is currently in progress.

19.3 RESULTS

At present, more than 40% of the land in the region is unfit for agricultural use (swamps, areas under water, tree and shrub lands, and other lands, including wind-

TABLE 19.1 Areal Changes in Land Use, Lagan Region, 1978–1996

Land-Use Category	Change in Area, 1978–1996 (ha)	1996 Area (% of 1978 Area)
Increasing or stable		
Submerged	+14,078	716.1
Marshes	+96,723	461.8
Roads	+5,196	418.2
Tree and brush lands	+10,320	396.0
Windblown sand	+2,092	118.4
Forests	+517	114.5
Others	+7,095	110.0
Decreasing		
Pastures	−73,556	76.8
Plowed land	−1,349	48.0
Meadows	−71,253	3.4

Source: Data calculated from remote survey data and materials from the State Committee on Land Resources of the Republic of Kalmykia.

TABLE 19.2 Major Land Users in Lagan Region and Land Areas (Hectares) as Reclassified after Flooding from January 1996 to January 1997

	Total	Agricultural Land			Submerged Land			Marshes			Miscellaneous		
		1996	1997	Change	1996	1997	Change	1996	1997	Change	1996	1997	Change
Dagestan Republic	158,013	118,350	108,135	−10,215	6,701	11,051	4,350	290	6,008	5,718	6,751	6,751	0
Jalykov Joint-Stock Company	71,971	59,673	41,901	−17,772	1,102	1,085	17	4,593	6,748	2,155	2,599	17,928	15,329
Artezianskiy State Farm	68,239	9,992	5,728	−4,264	26	26	0	56,269	61,993	5,724	38	11	−27
Krasinskiy State Farm	32,745	20,558	15,602	−4,956	676	723	47	8,805	14,360	5,555	611	564	−47
Kaspiyetz Collective Farm	18,326	9,219	9,884	665	523	569	46	6,999	7,083	84	599	610	11
Krasniy Moriak Collective Farm	16,589	7,464	6,196	−1,268	478	441	−37	6,100	8,661	2,561	408	374	−34
Rakushinskiy State Farm	11,741	2,120	60	−2,060	48	19	−29	9,382	11,655	2,273	24	—	−24
Lagan Executive[a]	11,700	4,466	4,486	20	1,084	1,084	0	3,950	3,950	0	478	480	2
Krasinskiy Rural Executive	2,092	1,136	673	−463	11	10	1	404	1,110	706	133	133	0
Total major land users	391,416	232,978	192,665	−40,313	10,649	15,008	4,395	96,792	121,298	24,776	11,641	26,851	15,210
Total Lagan region	468,551	288,627	247,354	−41,273	11,605	16,363	4,758	100,020	123,460	23,440	15,283	29,060	13,777

Source: Data from the State Committee on Land Resources of the Republic of Kalmykia.

[a]Lagan Executive had a total of 11,689 ha in 1996 and 11,711 ha in 1997.

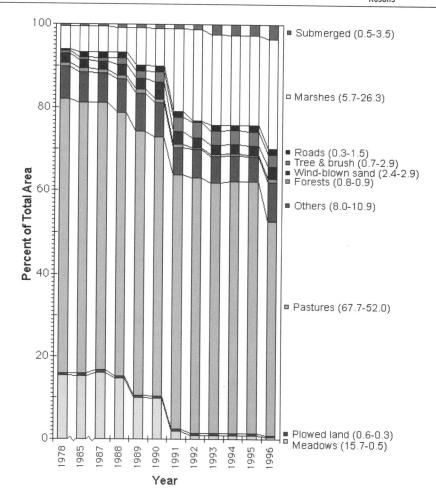

Figure 19.4 Land-use dynamics in the Lagan region. Bars indicate percent of total area; range in legend indicates percentage in 1978 and in 1996. (Data from the State Committee on Land Resources of the Republic of Kalmykia and the Committee on Land Resources of the Lagan Region.) "Meadows" includes the category "marshy meadows" used from 1978 to 1989; "pastures" includes the category "marshy pastures" used from 1978 to 1991; "forests" includes the category "forested lands" used from 1978 to 1989; "others" includes the categories "public courtyards, streets, squares," "public buildings," and "degraded lands."

blown sands). Figure 19.4 and Tables 19.1 and 19.2 clearly show the sharp increase in the proportion of these land categories as the sea level has risen.

Flooding and partial flooding have become widespread throughout the Lagan region. Virtually no farms, industrial installations, municipal facilities, or private homes remain unaffected. Flooding has resulted in major changes in land utilization, which have had negative economic and social consequences. The most drastic change has occurred in agricultural lands.

In 1996 a new land inventory was completed to update records of agricultural land lost to the rising Caspian and to record significant changes in land utilization.

A total of six townships have control over 29,763 ha of land (6.4% of the region), of which 17,539 ha is agricultural land (pasture only).

The overall picture of change in land use in the Lagan region, dating back to the beginning of Caspian sea-level rise, can be re-created from the data in Table 19.1. The continuous reduction in the area of productive agricultural lands and the increase in the percentage of idle lands are clearly visible, especially with regard to marshes and submerged areas.

19.3.1 Pastures

By early 1997 the land area of the region was 468,500 ha (Table 19.2) and the primary specialty was cattle farming. This is a reflection of regional physical and geographic properties (semiarid; hilly plains with xerophytic vegetation on brown soils). Until the late 1980s the region was a major hay source; hay was cut from a 15-km-wide area that became available as the sea receded. Those fields yielded from 3 to 6 Mt/ha of fodder. These changes can also be measured by the loss of 291 farms from 1978 to 1996 (78.9% of the 1978 farms were lost; same data sources as Table 19.1).

Table 19.2 shows land-use patterns for the principal government-owned, municipal, and cooperative land users in the Lagan region over the past two years. The two farms that were involved primarily in hay production, Artezianskiy State Farm (a farm in the Chernozemelskiy region) and Rakushinskiy State Farm, were established on lands belonging to the former State Land Authority under the auspices of the Central Department of the Black Lands and are located in the coastal area.

Pastures still account for the majority of lands (52%, Figure 19.4), but this percentage has declined since 1978 to only 76% of the 1978 area (Table 19.1). The reduction is caused by partial and complete flooding, and degradation has resulted from overgrazing, which is easy to follow on photographs along the entire coast (Figure 19.3).

From 1996 to 1997 the Republic of Kalmykia has seen pasture area of the farms reduced by 23,605 ha (18%). The area formally rezoned as nonagricultural land was 1018 ha, while 1414 ha was added to pastureland, including 322 ha of former plowed fields.

In contrast to the decline in pastures, the area of marshy pasture has grown exponentially. From 1978 to 1991, marshy pastures increased by 6038 ha, 2311% of the 1978 area (an increase by a factor of almost 22 over the preceding 13 years). Space photographs provide the best data for surveying this type of land.

19.3.2 Meadows

Since the beginning of the latest advance of the Caspian (approximately 1978) the region has lost almost all its meadows to flooding (Figure 19.4 and Table 19.1). The area covered by meadows has dropped from more than 73,700 ha in 1978 to less than 2500 ha, leaving only 3.4% of the original meadow area.

The connection between meadow loss and the rise in the sea level is further confirmed by the increased area of marshy meadows observed over the same period. As of 1992, statistical reports no longer mention "marshy meadows," which makes it expedient to use space photographs and field surveys when analyzing fodder field transformation over the last five years. From 1978 to 1989, marshy meadows increased by 5593 ha, to 3595% of their 1978 area.

Rising seas have flooded the coastal reed belt, several islands, and meadows. Changes in water balance have resulted in degradation of the unflooded coastal meadows and the replacement of meadow flora by the southern reed community (*editor's note*: probably *Phragmites australis*). These types of vegetation are easily identifiable from space photographs.

19.3.3 Plowed Land

Plowed fields accounted for no more than 0.6% of the entire area. This percentage has been more or less stable over the last 20 years, with an overall trend toward gradual dwindling (Figure 19.4). The region has no remaining irrigated land. As of early 1997 there was about 247,000 ha of agricultural land in the region (almost 53% of the total area, Table 19.2).

All the primary farms in the region have continued to see a reduction in the area of plowed lands. From 1996 to 1997, 932 ha of plowed land has been rezoned, 610 ha as nonagricultural land. The Kaspiyetz Collective Farm and the Krasinskiy State Farm have irrigated, plowed lands which have lost their economic value completely because of flooding by the Caspian and excessive moisture content.

Irrigated lands no longer exist, because of the rise in the water table. A land status report dated January 1, 1997 shows 1249 ha of irrigated land that is not usable due to (1) flooding, (2) excessive moisture content, (3) high water table, (4) degradation of the irrigation network or (5) lack of irrigation equipment. Some plowed land has been converted to pasture to compensate for grazing land that has been degraded.

19.3.4 Forests

Forests and forested lands are so few as to be insignificant and occupy no more than 0.9% of the total area (Figure 19.4), with only 0.2% forested lands. Multiple planting attempts were unsuccessful, primarily because of the rising water table and partial flooding by seawater. Most of the tree nurseries perished because the root systems were partially submerged or because entire areas were flooded.

19.3.5 Marshes and Submerged Lands

At present more than 40% of the region is unsuitable for use (marshes, submerged, tree and brush lands, and miscellaneous lands, including sands subject to wind erosion). Figure 19.4 and Table 19.1 illustrate the rise in percentages of these types of lands and the decline in meadows as sea level rises. At present, marshes extend in a wide belt along the entire coastline and even farther inland in the heart of the region (Figures 19.3 and 19.5). An increase in the areas of these land categories is directly attributable to the rise in the level of the Caspian Sea. According to statistical data,

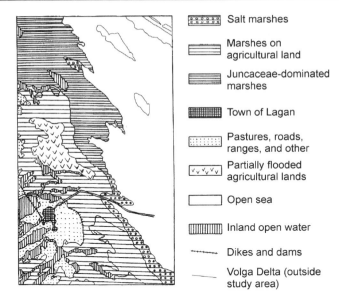

Figure 19.5 Land-use map of the Lagan coastal region, Kalmykia, based on interpretation of space photographs.

marshes and submerged lands currently occupy 29.8% of the area of the entire region, and the upward trend continues. Between 1996 and 1997, the area of the marshes grew by 23% while submerged areas increased by almost 41%. The area of extremely marshy lands amounts to 16,300 ha. This land category can easily be identified in the space photographs (Figure 19.3).

19.3.6 Soil Degradation

The Lagan region is part of the Black Lands and is characterized by rapid desertification, including especially severe soil deflation and secondary salinization. The region has 92,600 ha of saline soils and 125,200 ha of deflated soil, which equals 69 and 93.4%, respectively, of all Kalmyk agricultural lands. The areas affected by windblown sand have also increased dramatically since 1978 (Table 19.1).

19.3.7 Roads

The area of land used for roads is on the rise (Figure 19.4, Table 19.1). To a certain extent, this is caused by the construction of protective dams around the main townships in the coastal region (Lagan, Severnoye, Krasinskoye, Dzhalykovo, and Burannoye). The surfaces of earthen dams (the usual width of the upper part is 18 m) have been turned into roads. Most of these roads are built on embankments, which create conditions for the formation of a system of distinct *check plots* (enclosed flat plots without drainage). Within the check plots, active processes are marsh formation, rise in groundwater, and soil salinization. Protective structures, roads, and the system of check plots can be seen in more detail in aerial photographs (not shown), but the effect of building dams as protective installations that hinder inundation of the territories can be seen in space photographs (Figure 19.3).

19.3.8 Waste Depositories

Official numbers indicate a downward trend in the area of waste depositories and dumps caused by the shutdown of some "official" dumps (percentage of total area in 1992, 0.01; 1993 to 1995, 0.004; 1996, 0.003; data sources as for Figure 19.4). However, the number of unauthorized dumps and their area are on the increase. The rising seas of the Caspian caused dumps to the north of the city of Lagan to become partially submerged and completely washed away in the flood of 1995.

19.4 CONCLUSION

Rising seas have caused the flooding of a total of 150,000 ha of agricultural land and threaten to flood lands under development. The primary land-use change is the loss of agricultural land, principally the meadows that are the foundation of the feed base for cattle. Such changes have had an impact on the stability of cattle ranching, reflected in part by the declining number of cattle in the region. The best irrigated lands have completely lost their agricultural value. The area of nonproducing land (marshes and submerged lands) is increasing. In this context those areas that are taken up by Baer's mounds (linear ridges of probable aeolian origin) have become the most valuable land. They are islands to which facilities of the social and economic infrastructure can be relocated if the sea continues to advance.

An analysis of the space photographs in this study made it possible to show in detail the geographical distribution of the territories with changes in land use patterns and to detect the coastal areas of marshy lands and swamps (inaccessible for field studies; see Figures 19.3 and 19.5). Further processing of the photographs with a quantitative calculation of the areas of flooded and partially flooded lands will make it possible to better assess the losses in agricultural production resulting from a rise in the level of the Caspian Sea.

ACKNOWLEDGMENTS

I thank the following agencies for providing land statistics and other data for the Lagan region: Caspian Region subsidiary of the Priroda State Scientific Research and Production Center, Committee on Land Resources of the Lagan Region, Lagan Region Environmental Protection Committee, Republican Statistical Committee, and State Committee on Land Resources of the Republic of Kalmykia. I acknowledge TechTrans International, Inc. (Houston) for translating this manuscript into English, and staff members of the NASA Office of Earth Sciences for figure drafting and constructive reviews of the chapter.

Appendix

The Astronauts and Cosmonauts of the Shuttle-*Mir* Program

Figure A.1 NASA 2 and *Mir 21*, summer 1996 (NASA photograph NM21-395-024): crew members Shannon Lucid (left), Yuri Usachev, and Yuri Onufrienko.

Figure A.2 *Mir 22*, September 1996 (NASA photograph STS079-349-023): crew members Alexander Kalery (left), John Blaha (NASA 3), and Valery Korzun.

Figure A.3 NASA 4 and NASA 5, May 1997 (NASA ESC image S84E5028): crew members Jerry Linenger and Mike Foale exchange places on *Mir*.

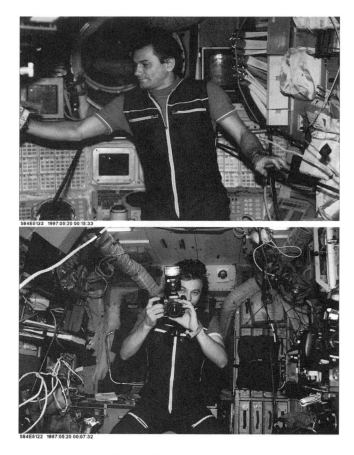

Figure A.4 *Mir 23,* May 1997 (NASA ESC images S84E5123 and S84E5122): (*A*) *Mir 23* commander Vasily Tsibliyev reaching for a camera; (*B*) *Mir 23* flight engineer Alexander (Sasha) Lazutkin, taking a picture.

Figure A.5 *Mir 24* and NASA 6, October 1997 (NASA ESC image 86E5391): *Mir 24* crew members Pavel Vinogradov (left) and Anatoly Soloviev (right), with NASA 6 crew member Dave Wolf in the middle.

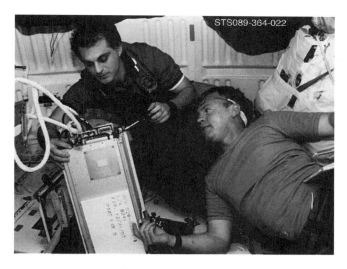

Figure A.6 NASA 6 and NASA 7, January 1998 (NASA photograph STS089-364-022): NASA-6 crew member Dave Wolf (left) works on some equipment with Andy Thomas (NASA 7).

Figure A.7 STS-60, February 1994. (NASA photograph STS060-31-028). The STS-60 crew was the first Russian–American crew since *Skylab*, 20 years earlier. Cosmonaut Sergei Krikalev (second from left) joined astronauts Franklin Chang-Dìaz, Jan Davis, Ron Sega (bottom row), and Ken Reightler and Charlie Bolden (top row) aboard the Space Shuttle. The mission objective was to rendezvous and fly around *Mir* without docking.

Figure A.8 STS-63, February 1995 (NASA photograph STS063-06-018). STS-63, the second Shuttle mission in the Phase I program, also flew to *Mir* for a nondocking rendezvous. On the left and right, in the front are Bernard Harris and Mike Foale, in the center are Janice Voss and Jim Wetherbee, and in the back are Russian cosmonaut Valdimir Titov and Eileen Collins.

Figure A.9 STS-71, June 1995 (NASA photograph STS071-122-013). The STS-71 mission was the first U.S. spacecraft to dock with the *Mir*. It retrieved the first U.S. crew member on *Mir*, Norm Thagard (upside down just right of center), and his crew mates Vladimir Dezhurov and Gennadi Strekalev (*Mir 18*). The crew members in this photo are, clockwise from center, Greg Harbaugh (dark shirt), Hoot Gibson, Charlie Precourt, Nikolai Budarin (*Mir 19*), Ellen Baker, Bonnie Dunbar, Norm Thagard, Vladimir Dezhurov, Gennadi Strekalev, and Anatoly Soloviev (*Mir 19,* in front with feet to the right).

Figure A.10 STS-74, November 1995 (NASA ESC image 743E18005). The second docking between the Shuttle and the *Mir* was STS-74. Although no crew members were exchanged, the STS-74 crew worked jointly with the *Mir 20* crew while docked. The crew are, clockwise from the top, Jerry Ross, Ken Cameron, Bill McArthur, Jim Halsell, and Chris Hadfield (Canadian Space Agency).

Figure A.11 STS-76, March 1996, (NASA photograph STS076-371-002). The STS-76 crew ferried Shannon Lucid to *Mir*. Front row, from left: Linda Godwin, Kevin Chilton, and Rick Searfoss. Back row: Rich Clifford, Shannon Lucid, and Ron Sega.

Figure A.12 STS-79 and *Mir 22*, September 1996 (NASA photograph STS079-349-022). The STS-79 and *Mir 22* crews, as Shannon Lucid ends her 6-month stay on *Mir*, and John Blaha begins the NASA 3 mission. Front row, from left: Alexander Kalery (*Mir 22*), Jay Apt, John Blaha (NASA 3), Bill Readdy, and Shannon Lucid (NASA 2). Back row: Tom Akers, Carl Walz, Valery Korzun (*Mir 22*), and Terry Wilcutt.

Figure A.13 STS-81, January 1997 (NASA ESC image S81E5535). The STS-81 crew brought Jerry Linenger up to *Mir* and John Blaha back to Earth. Front row, from left: Jeff Wisoff, John Blaha (NASA 3), Marsha Ivins, Alexander Kalery (*Mir 22*, in foreground), and Valery Korzun (*Mir 22*); center: Mike Baker; back: John Grunsfeld, Brent Jett (top), and Jerry Linenger (NASA 4).

Figure A.14 STS-84 and *Mir 23*, May 1997 (NASA photograph STS084-366-015). Crew members from *Mir 23* and STS-84 assemble for a group portrait. Front row, from left: Jerry M. Linenger (NASA 4), Vasili Tsibliyev (*Mir 23*), Charlie Precourt, Alexander Lazutkin (*Mir 23*), and Michael Foale (NASA 5). Back row: Ed Lu, Eileen Collins, Jean-François Clervoy, Elena Kondakova, and Carlos Noriega. Clervoy represents the European Space Agency (ESA) and Kondakova represents the Russian Space Agency (RSA).

Figure A.15 STS-86 and *Mir 24*, September 1997 (NASA photograph STS086-371-004). STS-86 crew members are joined by the *Mir 24* crew in the Spacehab Module aboard the Space Shuttle *Atlantis*. NASA 6 crew member David Wolf holds a cap, on the right side of the photo. Counterclockwise from Wolf are Vladimir Titov, Anatoly Soloviev (*Mir 24*), Scott Parazynski, Pavel Vinogradov (*Mir 24*), Jim Wetherbee, Wendy Lawrence, Mike Foale (NASA 5), Mike Bloomfield, and Jean-Loup Chrétien. Titov, from the Russian Space Agency (RSA), also flew on STS-63, the second predocking rendezvous mission to *Mir*. Chrétien represents the French Space Agency (CNES).

Figure A.16 STS-89 and *Mir 24,* January 1998 (NASA photograph STS089-391-004). Ten astronauts and cosmonauts form a human oval in order to fit into a single frame, onboard Russian *Mir* space station. Right-side up, from the left, are Dave Wolf, (NASA 6), Pavel Vinogradov (*Mir 24*), Terry Wilcutt, Anatoly Soloviev (*Mir 24*), and Bonnie Dunbar. Demonstrating the freedom of microgravity, head to head with bottom row, are (from the left) Salizhan Sharipov, Jim Reilly, and Joe Edwards, Jr. At 90° angle on the right are Andy Thomas, NASA 7 crew member (top); and Mike Anderson. Sharipov represents the Russian Space Agency (RSA).

Figure A.17 STS-91 and *Mir 25,* June 1998 (NASA photograph STS091-703-031). The STS-91 crew and the *Mir 25* cosmonauts pose for the final joint inflight Shuttle-*Mir* portrait in the *Mir* core module, at the end of Thomas' NASA 7 mission. Left to right are Valery Ryumin, Wendy Lawrence, Charlie Precourt, Andy Thomas, Talgat Musabayev (at center, *Mir 25*), Janet Kavandi, Dominic Gorie, Nikolai Budarin (*Mir 25*) and Franklin Chang-Diaz. Ryumin represents the Russian Space Agency (RSA).

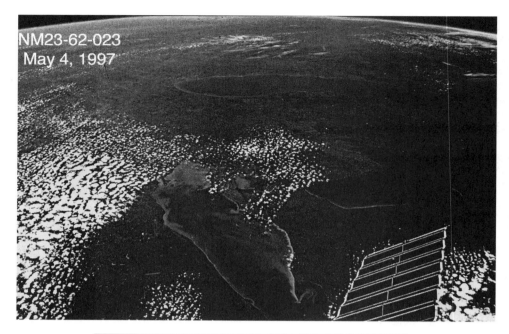

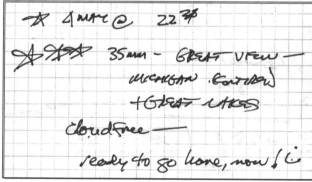

Figure A.18 NASA 4 crew member Jerry Linenger took this photograph (NASA photograph NM23-62-023) of the Great Lakes and his hometown in Michigan with thoughts of home near the end of his mission. Below he describes the scene in his notebook.

Index

A

Aerosol imaging, 77–98
   anthropogenic examples, 82–90
      China (Red Basin, Sichuan), 89
      Italy (northern), 86–87
      Nile River delta, 89–90
      southern Africa, 90–91
      Ukraine (western), 87–88
      United States (eastern), 83–86
   astronaut photographic documentation, 78–79
   desert dust, significance of, 81–82
   dust examples, 91–95
      Andean plumes, 92–93
      Chinese dust in Oregon, 93–95
      Sahara, 91
      Tibetan Plateau, 91–92
   industrial aerosols, significance of, 79–81
   overview, 77–78
Africa:
   aerosol imaging
      Sahara, 91
      southern, 90–91
   biomass burning, south-central, 112–115
   El Niño (1997–1998), 71–72
   Okavango Swamp (Botswana), 54–56

African Sahel, Niger River delta (inland), 54
Agricultural land, Caspian Sea coastal land-use changes, 237
Andean plumes, aerosol imaging, 92–93
Anthropogenic haze phenomena. *See* Aerosol imaging
Aral' Sea, water level fluctuations, 47–48
Armistad Resevoir (Texas-Mexico), water level fluctuations, 48–49
Asia. *See also* Caspian Sea
   Central, Aral' Sea, 47–48
   China, biomass burning, 109–110
   Southeast
      biomass burning, 110–111
      El Niño (1997–1998), 69–70
Atmospheric aerosols. *See* Aerosol imaging
Australia, El Niño (1997–1998), 71
Azerbaijan. *See* Caspian Sea coastal responses

B

Base Block windows, 124–126
Bermuda High, aerosol imaging, 83–86

Biomass burning, 99–119
　examples, 108–115
　　Mexico-Central America, 111–112
　　Russian Far East, China, and Mongolia, 109–110
　　south-central Africa, 112–115
　　Southeast Asia, 110–111
　methods, 101–102
　overview, 99–101
　patterns, 102–108
　　geographical distribution, 103–105
　　　lower latitudes, 104–105
　　　northern latitudes, 104
　　seasonal distribution, 105–108
　　　lower latitudes, 107–108
　　　northern latitudes, 106
Botswana, Okavango Swamp, 54–56

C

Caspian Sea. *See also* Gulf of Kara-Bogaz-Gol; Volga River delta
　eddy formation, 211–219
　　currents, 212
　　overview, 211
　　surface winds and internal waves, 212–216
　　upwelling water, 216–218
　environmental problems, 140
　Gulf of Kara-Bogaz-Gol, 201–210 (*See also* Gulf of Kara-Bogaz-Gol)
　investigation results, 140–141
　northern coast, 149–157
　　deep-seated structures, 150–155
　　overview, 149
　　sedimentary cover, 155–156
　physical characteristics, 133–137
　　climate, 135–136
　　geography, 133–135
　　salinity, 136–137
　　water balance, 137
　regional economy, 137–138
　sea-level fluctuations, 138–140, 145–148

　avian habitats, 171–180 (*See also* Volga River delta (avian habitats))
　coastal zone dynamics, 191–199 (*See also* Caspian Sea coastal zone dynamics)
　vegetation changes, 181–190 (*See also* Caspian Sea coastal vegetation)
Caspian Sea coastal land-use changes, 231–239
　methods, 232–233
　overview, 231–232
　results, 233–239
　　forests, 237
　　generally, 233–236
　　marshes and submerged lands, 237–238
　　meadows, 236–237
　　pastures, 236
　　plowed land, 237
　　roads, 238–239
　　soil degradation, 238
　　waste depositories, 239
Caspian Sea coastal responses, 221–229
　geological/geomorphological characteristics, 222–226
　　generally, 222–223
　　Gobustan foothills, 225–226
　　Kura depression, 223–225
　　Talish Mountains, 226
　modern characteristics, 226–228
　　Gobustan coastline, 228
　　Kura River delta, 226–227
　　Kura spit, 227–228
　overview, 221–222
Caspian Sea coastal vegetation, 181–190
　1990's, 185–188
　overview, 181–182
　regression period, 182–184
　sea-level rise, 184–185
　succession, 188
Caspian Sea coastal zone dynamics, 191–199
　overview, 191–192

regression stage, 1976, 192–193
transgression stage
   1985, 193–197
   1990's, 193–197
Central America, biomass burning, 111–112
Central Asia, Aral' Sea, 47–48
China:
   aerosol imaging
      Chinese dust in Oregon, 93–95
      Red Basin, Sichuan, 89
      Tibetan Plateau, 91–92
   biomass burning, 109–110
Climate, Caspian Sea, 135–136
Climate change:
   aerosol imaging, 77–98 (*See also* Aerosol imaging)
   El Niño (1997–1998), 61–76 (*See also* El Niño (1997–1998))
   water level fluctuations, 43–60 (*See also* Water-level fluctuations)
Coastal vegetation, Caspian Sea, 181–190. *See also* Caspian Sea coastal vegetation
Currents, eddy formation, Caspian Sea, 212

D
Desert dust, significance of, aerosol imaging, 81–82. *See also* Aerosol imaging
Desertification, Caspian Sea coastal land-use changes, 238
Drought, El Niño (1997–1998), 61–76. *See also* El Niño (1997–1998)
Dumps, Caspian Sea coastal land-use changes, 239
Dust, significance of, aerosol imaging, 81–82. *See also* Aerosol imaging

E
Egypt:
   Lake Nasser, 50–52
   Nile River delta, 89–90

El Niño (1982), Andean plumes, aerosol imaging, 92–93
El Niño (1997–1998), 61–76
   Africa, 71–72
   Australia, 71
   background, 62–65
   biomass burning, 100, 111 (*See also* Biomass burning)
   North America, 72–73
   overview, 61–62
   South America, 66–69
   Southeast Asia, 69–70
ENSO. *See* El Niño
Euphrates River, Greater Anatolia Project, 52–54

F
Floods:
   El Niño (1997–1998), 61–76 (*See also* El Niño (1997–1998))
   Ohio-Mississippi Rivers, 49–50
Forests, Caspian Sea coastal land-use changes, 237

G
Global warming. *See* Climate change
Gobustan coastline, Caspian Sea coastal responses, 228
Gobustan foothills, Caspian Sea coastal responses, 225–226
Greater Anatolia Project (Turkey), water level fluctuations, 52–54
Gulf of Kara-Bogaz-Gol, 201–210
   geomorphological changes, 203–207
   natural conditions, 202–203
   overview, 201–202
   present conditions, 207–209

I
Ice fields (Patagonia), water level fluctuations, 56, 57
Indonesia, biomass burning, 110–111

Industrial aerosols, significance of, aerosol imaging, 79–81. *See also* Aerosol imaging
Italy, aerosol imaging, 86–87

K

Kura depression, Caspian Sea coastal responses, 223–225
Kura River delta, Caspian Sea coastal responses, 226–227
Kura spit, Caspian Sea coastal responses, 227–228
Kvant-2 window, 127–128

L

Lake Armistad (Texas-Mexico), water level fluctuations, 48–49
Lake Nasser (Egypt), water level fluctuations, 50–52
Littoral upwelling water, eddy formation, Caspian Sea, 216–218

M

Marshes, Caspian Sea coastal land-use changes, 237–238
Meadows, Caspian Sea coastal land-use changes, 236–237
Mediterranean Sea, Saharan dust, aerosol imaging, 91
Mexico:
 Armistad Resevoir, 48–49
 biomass burning, 111–112
 Mississippi-Ohio Rivers, water level fluctuations, 49–50
Mongolia, biomass burning, 109–110

N

Niger River delta (inland), water level fluctuations, 54
Nile River, Lake Nasser (Egypt), 50–52
Nile River delta, aerosol imaging, 89–90
North America:
 aerosol imaging, 83–86

 El Niño (1997–1998), 72–73
 Mississippi-Ohio Rivers, 49–50
 urban growth in (*See* Urban growth (North America))

O

Ohio-Mississippi Rivers, water level fluctuations, 49–50
Okavango Swamp (Botswana), water level fluctuations, 54–56
Optical windows. *See* Windows
Oregon, aerosol imaging, Chinese dust in, 93–95

P

Pastures, Caspian Sea coastal land-use changes, 236
Patagonia, ice fields, water level fluctuations, 56, 57
Plowed land, Caspian Sea coastal land-use changes, 237
Pollution:
 aerosol imaging, 77–98 (*See also* Aerosol imaging)
 Caspian Sea, 140
Population size, remote sensing, 26–27. *See also* Urban growth
Precipitation:
 Caspian Sea, 135–136
 El Niño (1997–1998), 61–76 (*See also* El Niño (1997–1998))
 water level fluctuations, 44, 45
Priroda window, 128

R

Red Basin, Sichuan, China, aerosol imaging, 89
Remote sensing, urban growth (North America), 26–27. *See also* Urban growth
Rio Grande, Armistad Resevoir, 48–49
Roads, Caspian Sea coastal land-use changes, 238–239
Russian Far East, biomass burning, 109–110

Russian visual observations, 15–23
  operational results, 19–22
  overview, 15–17
  training, 18–19

S

Sahara:
  aerosol imaging, 91
  Niger River delta (inland), 54
Salinity, Caspian Sea, 136–137
Salyut 6. See Russian visual observations
Sea-level fluctuations, Caspian Sea, 138–140, 145–148. See also Caspian Sea; Water-level fluctuations
Sedimentary cover, Caspian Sea, 155–156
Shoreline dynamics, Volga River delta, 161–166. See also Volga River delta
Shuttle-*MIR* earth science investigations, 1–14
  goals of, 1–2
  historical perspective on, 2–3
  methods, 3
  results, 3–13
    generally, 3–5
    of missions, 5–12
    of preliminary science, 12–13
  windows for, 122–129 (See also Windows)
Sichuan, China, aerosol imaging, 89
Smoke palls. See Biomass burning
Snowmelt, ice fields (Patagonia), water level fluctuations, 56, 57
Soil degradation, Caspian Sea coastal land-use changes, 238
South America:
  Andean plumes, aerosol imaging, 92–93
  El Niño (1997–1998), 66–69
South-central Africa, biomass burning, 112–115
Southeast Asia:
  biomass burning, 110–111
  El Niño (1997–1998), 69–70
Southern Africa, aerosol imaging, 90–91
Submerged lands, Caspian Sea coastal land-use changes, 237–238
Surface winds, eddy formation, Caspian Sea, 212–216

T

Talish Mountains, Caspian Sea coastal responses, 226
Texas, Armistad Resevoir, 48–49
Tibetan Plateau, aerosol imaging, 91–92
Tigris River, Greater Anatolia Project, water level fluctuations, 52–54
Turkey, Greater Anatolia Project, water level fluctuations, 52–54

U

Ukraine, aerosol imaging, 87–88
United States:
  aerosol imaging, 83–86
  Mississippi-Ohio Rivers, 49–50
Upwelling water, eddy formation, Caspian Sea, 216–218
Ural River delta, Caspian Sea, 151, 152, 153, 154
Urban growth (North America), 25–41
  methods, 28–32
  objectives, 27–28
  overview, 25–26
  remote sensing, 26–27
  results, 32–38

V

Vegetation, Caspian Sea, 181–190. See also Caspian Sea coastal vegetation
Visual Observations of the Earth. See Russian visual observations; Shuttle-*MIR* earth science investigations
Volga River delta, 152, 153, 154. See also Caspian Sea

Volga River delta (avian habitats), 171–180
  methods, 172–174
  overview, 171–172
  results, 174–179
    high sea level, 178–179
    low sea level, 174–175
    sea-level rise, 175–178
Volga River delta (shoreline dynamics and hydrographic system), 159–169
  methods, 160
  overview, 159–160
  results, 160–167
    channel evolution, 166–167
    shoreline dynamics, 161–166

W

Waste depositories, Caspian Sea coastal land-use changes, 239
Water balance, Caspian Sea, 137
Water-level fluctuations, 43–60. *See also* Caspian Sea; El Niño (1997–1998)
  El Niño, 61–76
  methods, 46–47
  overview, 43–46
  regions, 47–56
    Aral' Sea, 47–48
    Armistad Resevoir (Texas-Mexico), 48–49
    Greater Anatolia Project (Turkey), 52–54
    ice fields (Patagonia), 56, 57
    Lake Nasser, 50–52
    Niger River delta (inland), 54
    Ohio-Mississippi Rivers, 49–50
    Okavango Swamp (Botswana), 54–56
Waves, eddy formation, Caspian Sea, 212–216
Wetlands. *See* Volga River delta (avian habitats)
Windows, 121–129
  methods, 122–123
    image data analysis, 123
    on-orbit image acquisition, 122
  overview, 121–122
  results, 124–128
    Base Block windows, 124–126
    Kvant-2 window, 127–128
    Priroda window, 128
Winds, eddy formation, Caspian Sea, 212–216

Photo Insert Captions

Figure 3.1 (*A*) The Greater Chicago area as photographed in 1973 and 1996 (NASA photograph SL3-46-199 and NM21-767-068). Small arrows indicate Merrill C. Meigs Field used for scale. (*B*) Dallas and Fort Worth in 1969 and 1996 (NASA photograph AS9-21-3299 and NM22-774-019). Small arrows indicate the Dallas/Fort Worth International Airport used for scale. (*C*) Las Vegas in 1973 and 1996 (NASA photograph SL3-28-59 and NM22-725-034). Small arrows indicate McCarran International Airport used for scale.

Figure 3.5 Built-up areas identified in 1969 or 1974 photographs (yellow) and 1996 photographs (red) overlaid on the registered and resampled image from NASA-*Mir* (1996): (*A*) San Francisco Bay area; (*B*) Mexico City; (*C*) Vancouver; (*D*) Dallas/Fort Worth; (*E*) Chicago; (*F*) Las Vegas.

Figure 4.2 Aral' Sea. This set of four photographs (NASA photographs STS51F-36-059, STS047-79-082, NM21-762-025A, and NASA6-707-034) provides a pictorial time series of water-level drops in the Aral' Sea from 1985 through early 1998. Note that the sequence includes the cutoff near the Syr Darya delta (a) between the northern and southern basins between 1985 and 1992. Also note the shape of the Amu Darya deltaic coastline (b) and the progressively increased size of Vozrozhdeniya Island (c). Finally, between 1996 and 1998, Barsakel'mes Island, the arrow-shaped island (d), joins with the mainland to become a peninsula.

Figure 4.5 Floods in the Ohio and Mississippi Rivers. Top right: Ohio River flooding at Evansville, Indiana on March 9, 1997 (NASA photograph NM23-705-321). The feature on the left side is a *Mir* solar array. Top left: Flooding from the confluence of the Ohio and Mississippi Rivers to Memphis, Tennessee on March 15, 1998 (NASA photograph NM23-712-438). The area inside the white box, which includes the city of Memphis, is given in detail below. Bottom left: Detailed view of the area inside the box (NASA photograph NM23-712-438). The extent of flooding around Memphis is obvious. Bottom right: For comparison, image showing the same area during normal river flow in September 1996 (NASA photograph STS079-812-086).

Figure 4.8 Inland delta of the Niger River. (*A*) In October 1996 (NASA photograph NM22-736-055) the dark green vegetation is lush when water from the summer rains pass through the region. Although vegetation cover lasts only a few months, it provides a strong visual contrast with the surrounding desert countryside. (*B*) By April 1997, the wetlands have dried up, and the inland delta is a uniform tan color (NASA photograph NM23-730-271). The white box outlines the area covered by view (*A*).

Figure 5.1 Sea surface temperature anomaly map, December 8, 1997. (From Fleet Numerical Meteorology and Oceanography Center, 1998.) Precipitation anomalies estimated to be within the driest or wettest 10% of climatological occurrence between November 1997 and January 1998 have been plotted on the map. (Modified from Climate Prediction Center, 1998.) Regions affected by El Niño weather that are discussed in this chapter are marked by arrows connected to the text boxes. The temperature scale is in degrees Celsius. Mission durations are: NASA 5 from May to September 1997; NASA 6 from September 1997 to January 1998; and NASA 7 from January 1998 to June 1998.

Figure 5.4 This series of photos of Lago Poopó shows how rapidly the lake dried up during the 1997–1998 El Niño. This shallow lake in the Bolivian Andes is very responsive to climate shifts. The extreme fluctuations in water levels can be used as a visual indicator for relative rainfall in the region. The view in the upper left (NASA photograph NM23-714-627) was taken in March 1997 after unusually wet weather in the high Andes—up to 200% of normal—had flooded Lake Poopó (arrow) and nearby salars Uyuni and Coipasa (just right of center) for the first time in years. The photo in the upper right (NASA photograph NASA5-705-085) shows Lago Poopó in late July 1997. By November 1997 the "thumb" on Poopó had completely evaporated (lower left, NASA photograph NASA6-710-082). The lower right photo, taken in May 1998, completes the time series and shows that the lake is almost completely dry (NASA photograph NASA7-726-036). A chart showing regional cumulative precipitation over this time period is provided in Figure 5.3.

Figure 5.7 Lake Eyre in south-central Australia responds to El Niño events by drying up. These photos of Lake Eyre were taken in July 1997 (NASA photograph STS094-748-083), January 1998 (NASA photograph STS089-717-055), and April 1998 (NASA photograph NASA7-714-014) and depict the dropping water levels in Lake Eyre due to the regional, El Niño–induced drought. The graph shows the regional precipitation deviations due to El Niño. The cumulative observed precipitation is given by the heavy solid line, and normal precipitation (cumulative) is depicted by the thin dashed line. The red bars indicate when the photos were taken. (Modified from the Climate Prediction Center, 1998.)

Figure 5.8 Coastal Somalia in March 1996 and January 1998. The 1998 view (NASA photograph NASA6-708-056), taken after the heavy rains, shows remarkably green vegetation over this normally arid region. It contrasts with the brown, more normal view, taken in March 1996 (NASA photograph NM21-727-006). The large light-colored patches on the coast are dune fields. The graph shows the regional precipitation deviations due to El Niño. The cumulative observed precipitation is given by the heavy solid line, and normal precipitation (cumulative) is depicted by the thin dashed line. The bar indicates when the NASA 6 photo was taken. (Modified from the Climate Prediction Center, 1998.)

Figure 5.9 Heavy rains from December through April along the central California coast supported significant growth of the regional vegetation, which gave this section of California an unusually green cast. This image (NASA photograph NASA7-709-029) taken in March 1998, centers on San Francisco Bay. The urban region around the bay contrasts strongly with the green vegetation on the surrounding hillsides. The graph shows the regional precipitation deviations due to El Niño. The cumulative observed precipitation is given by the heavy solid line, and normal precipitation (cumulative) is depicted by the thin dashed line. The red bar indicates when the photo was taken. (Modified from the Climate Prediction Center, 1998.)

Figure 6.3 This Space Shuttle photograph shows a major haze event in the eastern United States on April 26, 1990, 14:49:03 GMT. The view looks obliquely north along the east coast from a point above the Caribbean Sea. A mass of aerosol haze stretches across the top of the entire view. This mass was transported west to east (left to right in this view) around the north limb of the high-pressure cell moving offshore for at least 1500 km, beyond the Atlantic islands of Bermuda (B). The leading edge of the haze mass can be detected (right center) north of the Bahamas (islands surrounded by light-blue seabed, bottom right), indicating that aerosols from the industrial northeast were transported around the high, with a final trajectory leading directly toward the large population centers of Florida. (NASA photograph STS031-151-155, center point 26°N 80°W, craft nadir 20.9°N 83.4°W, Linhof camera, 90-mm lens, altitude 617 km.)

Figure 6.8 Industrial haze flowing from the Po River Valley over the Adriatic Sea. Two panoramic views to the south-southwest, taken on successive days, show the Adriatic Sea (center left) and all of the peninsula of Italy (about 1000 km long), looking south from a point over the Alps whose rugged mountainous landscape appears in the foreground of view (*A*). The north end of the Adriatic Sea is partly obscured by industrial haze from the Po River valley (*A*, center and center right; *B*, foreground). The polluted air contrasts with the clearer air over the Adriatic farther south (middle ground)—and also contrasts with patches of cloud and snow on the Alps, both of which are brilliant white and well delineated. Arrows in view (*A*) indicate the north–south line separating denser from less dense industrial aerosols on October 25, 1997 (11:58:07 GMT); the next day (October 26, 1997, 10:59:51 GMT) this line had swung around to an east-west position (arrows, *B*) as the haze drifted slowly south down the channel of the Adriatic Sea. [NASA photographs, Hasselblad camera, 100 mm lens, altitude 383 km: (*A*) NASA6-704-83, center point 43.5°N 13°E, craft nadir 50.6°N 10.6°E; (*B*) NASA6-707-65, center point 42.5°N 15°E, craft nadir 48.9°N 10.7°E.]

Figure 6.10 Anthropogenic haze from China over the Pacific Ocean, early March 1996. (*A*) The entire Red Basin of Sichuan Province filled with gray anthropogenic haze on a winter day as anticyclonic conditions were developing. In this west-looking view, the Tibetan Plateau stretches across the top of the view, and the snowcapped Himalaya Mountains appear in the extreme top left corner. The distance across the Red Basin is approximately 450 km; the horizontal distance from the Space Shuttle nadir to the basin margin is also about 450 km. (*B*) A coherent corridor of anthropogenic haze (arrows, probably a mixture of industrial air pollution, dust, and smoke) can be seen in the left half of the view against the dark background of the East China Sea. The corridor is about 200 km wide and probably much longer than 600 km (visible distance over the sea). In this southwest-looking view, the island of Taiwan (T) appears top left (350 km in length) and the east coast of China across the rest of the view. Topographic detail in China is degraded under the thick layer of haze. This picture was taken as the Space Shuttle flew over Okinawa—the distance to Shanghai (at the near point on the Chinese coastline, top right) is about 650 km. [NASA photographs, Hasselblad camera, 40-mm lens: (*A*) STS075-721-22, between February 22 and March 9, 1996, center point 27°N 105°E, altitude 296 km; (*B*) STS075-773-66, March 4, 1996, 01:29:47 GMT, center point 28°N 123°E, craft nadir 28°N 128.1°W, altitude 278 km.]

Figure 6.11 Photomosaic of haze over the Nile River delta. Cairo, at the apex of the Nile delta in the lower half of the mosaic, is invisible beneath a blanket of haze. Under clear weather conditions, the city is visible (inset: arrows indicate north and south margins of city). Two white smoke plumes rise from the region of Helwan, south of Cairo, indicating that the ambient wind is from the west-northwest (top left toward bottom right) on this day. The regular, wavy structure on the surface of the pollution haze (top right) is transverse to the wind direction and appears to be a sequence of "gravity waves," features commonly developed between air layers of differing density (some waves are capped by small cumulus clouds, top right). South of the delta, removed from sources of the haze, the green floor of the Nile valley and the Faiyum depression (bottom left) appear distinctly clearer. [Mosaic of NASA photographs taken August 5, 1997, Hasselblad camera, 250-mm lens, altitude 383 km; NASA5-707-53 (upper), 15:39:01 GMT, center point 30.5°N 31°E, craft nadir 30.5°N 36.0°E; NASA5-707-52 (lower), 15:38:46 GMT, center point 30°N 31°E, craft nadir 29.8°N 35.2°E. Inset STS084-310-36, May 22, 1997, 05:49:49 GMT, center point 45°N 10°E, craft nadir 45.9°N 14.5°E, Hasselblad camera, 100-mm lens, altitude 372 km.]

Figure 6.14 Saharan dust plume moving into the Mediterranean basin, August 8, 1998. The red arrow marks the same location and cloud mass on each image. The progression from oblique to vertical look angles shows different aspects of the aerosols. (*A*) Northeast-looking panorama shows the landmasses of Spain lower left (city of Almeria, left arrow) and North Africa lower right (city of Tlemcen, Algeria, right arrow). Dust obscures the Balearic Islands (Palma city on the island of Mallorca, P, and the island of Ibiza, I). The Sahara Desert lies out of the picture, right. (*B*) Closer view of cloud and dust mass in view (*A*). The cloud mass has several embedded convection cells (center right) with apparently associated linear dust features at lower altitudes along its western margin (center, and left of red arrow). The island of Mallorca (P) is dimly visible in this more vertical view but the island of Menorca (M, 120 km east-northeast of Mallorca) remains obscured. (*C*) Detail of linear structure in the dust immediately beneath the edge of the largest mass of cloud [center of view (*B*)] suggests the influence of near-surface air outflow generated by downdrafts in the storm cells. In this vertical view taken through a thinner dust column, the island of Menorca (*M*) is partly visible. [NASA photographs taken August 8, 1997, Hasselblad camera, 100-mm lens, altitude 382 km: (*A*) NASA5-708-48, 17:29:07 GMT, center point 36.5°N 1.5°W, craft nadir 33.7°N 6.0°W; (*B*) NASA5-708-60, 17:30:54 GMT, center point 38°N 3°E, craft nadir 38.2°N 0.2°E; (*C*) NASA5-708-66, 17:31:37 GMT, center point 40°N 4.5°E, craft nadir 39.9°N 2.9°E.]

Figure 6.16 A set of brown dust plumes over Tibet on February 12, 1997 (11:06:11 GMT), start near the Paikü Co lake (dark blue, center), where winds are mobilizing ancient lake-bottom sediments. Whiter dust plumes appear top right. The alignment of the mountain ranges controls the trajectories of the plumes. This west-looking panorama shows 700 km of snowcapped peaks of the Himalaya Mountains (down the left side of the view) and the high, cold desert plateau of Tibet in the rest of the view. The upper Brahmaputra River occupies the major valley between the dust clouds (arrows). Clouds (lower left) and smoke haze (upper left corner) appear over lower-lying valleys of Nepal which lead down to the Ganges plain. (NASA photograph NM22-759-329, center point 29°N 86°E, Hasselblad camera, 100-mm lens, altitude 376 km.)

Figure 6.17 Plumes of mainly light-brown dust can be seen streaming off the high, arid plains of the central Andes Mountains in this view, which looks southwest toward the Pacific Ocean and Atacama Desert of northern Chile (across the top of view). Dust plumes are injected into the westerly winds above altitudes of 4200 m—this oblique view shows the main plume intersecting the dark Cachi massif (C) at about 6000 m. The visible length of the major dust plume in this view is approximately 240 km. The complexity of airstream circulations is apparent: dust plumes are transported by westerly winds (flow from top right to lower left) above altitudes of 4200 m. The stratus cloud, with an upper surface at 3290 m, is transported by gentle southerly winds (flow from left to right). The dust itself therefore rains down onto the stratus cloud (in the region of the arrows) before finally reaching the ground. This view shows that the dust pall increases albedos over the dark land surface, but decreases albedo of the cloud. (NASA photograph STS008-46-936, August 30, 1983, center point 25.5°S 66.5°W, Hasselblad camera, 100-mm lens; altitude 311 km.)

Figure 7.4 Fires in Mongolia, 1996. (A) On May 2, 1996, dense smoke rises from several large fires that raged out of control for more than 3 weeks. A dying fire and associated black fire scar appear at the top right of the view (NASA photograph NM21-735-62). (B) Scattered fires appear in this east-looking view of the mountainous and forested landscape on the border between western Mongolia and the Russian Federation, May 11, 1996. A small sector of Lake Baikal appears on the left edge of the view (NASA photograph NM21-743-56). Two centers of biomass burning, one near the west end of Lake Baikal (lower left) and one south of the lake (center) appear as diffuse gray masses (the distance from the lake to the lower-left fires is about 150 km). Smoke appears less brilliant than the lines of cloud which dominate the top and lower right parts of the view. Westerly winds carry smoke from fires burning on higher forested ground (green). Lower-lying country is a semiarid steppe and appears as a gray-brown.

Figure 7.5 (A) Fires on the Russian–Mongolian border, May, 1, 1997. This oblique, west-looking view of the forested mountain ranges in northern Mongolia and southern Russia (NASA photograph NM23-756-448) shows the same forested regions as in Figure 7.4. Lake Baikal appears at the top of the view still covered with areas of winter ice. Multiple, discrete point sources of smoke (lower left) originate within the mountainous country on the border between the Russian Federation and Mongolia. The fires appear to be located within but near the edges of the green, forested areas (upper inset, a zoomed view of fires in the region just below center), which occupy higher elevations. Smaller patches of smoke appear beyond the main forested region (top left). Winds take smoke southward (toward the lower left). At the bottom of the view are the semiarid steppes of northern Mongolia, which appear as a light brown. A prominent, angular burn scar can be seen, indicating a grassland fire (center foreground and lower inset). (B) Fires in China and the Russian Far East, May 2, 1997. About 700 km of the coastal ranges of Russia and the mountainous north of North Korea appear as the darker feature crossing the photograph (NASA photograph NM23-763-613). Most of the view shows a widespread smoke pall, both inland in China (top in this west-looking view) and on the seaward side of the mountains (bottom). The smoke pall area on May 2, 1997 can be estimated from this view to be at least half a million km^2. Individual fires appear on the seaward slopes with smoke plumes driven toward Japan by westerly winds (top right to lower left).

Figure 7.6 This mosaic of two electronic still camera images (NASA ESC images S86E5097 and S86E5098) shows an increase in smoke density in southern Sumatra, Indonesia, on September 26, 1997, during the ENSO-related smoke crisis in Southeast Asia. Forested slopes cover the flanks of many volcanoes in the south of Sumatra; farther north (left) forest fires can be seen as white wedge shapes that fan out from the fire source. In the north, a general pall of smoke obscures all the lower elevations of the landscape, leaving only the volcano summits protruding (lower left). Small masses of cloud appear as brilliant white patches. Accompanying data from the Total Ozone Mapping Spectrometer (TOMS) satellite data show contours of the aerosol index, a relative measure of the ultraviolet (UV) albedo for different wavelengths in the UV spectrum. The white arrows indicate the approximate positions of the top and bottom of the mosaic, right on the boundary between clear air and the smoke pall. (Modified from National Aeronautics and Space Administration, 1998a.)

Figure 7.7 Comparison of *Mir* photographs and satellite data on smoke from the Mexico–Central American fires of 1998. Contour maps of data from the TOMS satellite show the aerosol index, a relative measure of the ultraviolet albedo for different wavelengths in the UV spectrum. The tips of the white arrows indicate the approximate nadir location of the *Mir* and the arrow directions indicate the approximate look direction of the photographs. (*A*) Smoke palls in Mexico, May 16, 1998 (NASA photograph NASA7-726-20). The mountainous spine of the Sierra Madre del Sur in southern Mexico, 600 km long, runs diagonally across this southeast-looking view. The regional extent of a massive smoke pall is apparent on the Pacific flank of the Sierra, with the coastline dimly visible under the smoke (far right). Winds blow smoke plumes (long white tendrils) westward (to the right) from individual fires in forests at the higher elevations of the mountains. A mass of cloud (white region cutting across the view below the horizon) occupies the top of the view. (*B*) Central American smoke over Florida, May 19, 1998 (NASA photograph NASA7-725-22). This photograph was taken when the Mexico–Central America smoke transport reached its peak. This northeast-looking view shows smoke in the eastern Gulf of Mexico (foreground), partly obscuring the peninsula of Florida (middle ground), and extending hundreds of kilometers into the Atlantic Ocean. The rounded object in the lower right corner is a part of the *Mir* station. To aid interpretation, a sketch map of the photograph is provided in the lower left. (Modified from National Aeronautics and Space Administration, 1998b.)

Figure 8.1 The *Mir* complex as seen from the Space Shuttle (70-mm film, NASA photograph STS091-727-55, June 1998).

Figure 10.3 Ural River delta: NASA photographs (*A*) October 1994 (STS068-202-069); (*B*) May 1996 (NM21-740-055). The Caspian has also encroached upon the delta of the Ural River, the headwaters of which are in the southern Ural Mountains. Although it can be difficult to distinguish water from vegetation, there are differences (arrows). The most notable differences are in the southernmost lobe—for example, the widths of channels separating subaerial islands. In view (*B*) sunglint lends emphasis to submerged areas and one can see that exposed dredge spoil east of the main channel has decreased markedly (C). Along the shorter, western canal the banks that were exposed in 1994 were almost entirely submerged or eroded two years later (at A). Sunglint also highlights what appears to be a network of levees—possibly roads or canals—that have been flooded east of the river channel (at B).

Figure 10.5 Gulf of Kara-Bogaz-Gol, Turkmenistan: NASA photographs (*A*) April 1985 (STS61A-200-034); (*B*) April 1994 (STS059-L17-073); (*C*) January 1997 (NM22-735-050). A Turkmen legend holds that the Caspian separated from her husband, the Black Sea, which brought down a curse from Allah upon the Caspian. Their offspring, Kara Bogaz, would never cut its umbilical cord with its mother, and the Caspian would forever have to feed water to the bay (St. George, 1974). (*A*) In 1985, not long after flow between the Caspian and Gulf of Kara-Bogaz-Gol was partially reestablished, there was still little water in the bay; bright, reflective salt deposits occupied much of the area. Note the peninsulas along the southern bay margin (arrow). (*B*) A channel, visible in this southwestward view, was dredged through the spit in 1992 and rising Caspian waters spread over the bay floor (see Chapter 16). The spits and peninsulas along the south coast were significantly smaller in 1994. (*C*) By January 1997, when this detailed view was taken, the island north of the largest peninsula had virtually disappeared and much of the peninsula had been inundated (arrow). Of the two anvil-shaped promontories along the southern shore in the 1994 view, one has become an island and the other has lost the eastern point.

Figure 11.6 Geological and geomorphological interpretation of the Selitryanoe/Seroglazovka area of the Volga River valley, September 1994 (NASA photograph STS064-101-20). 1, Upper Pleistocene (Khvalyn) deposits of the plain; 2, riverbed and fluvial sediments; 3, modern beach accretion, sand bars; 4, lower floodplain and its deposits; 5, upper floodplain and its deposits; 6, oxbow lakes and their deposits.

Figure 13.4 Vegetation maps of southern portion of Damchik section of Astrakhanskiy Biosphere Reserve in 1977 and 1992. Key to vegetation types: E3, willow forests (*Salix alba* and *S. triandra*) with layer of common reed (*Phragmites australis*), reedmace (*Typha angustifolia*), forb, and grass; E4, open forests of *S. alba* with layer of sedge, forb, and grass; E17, reed meadows; A1, dense stand of reedmace with small amounts of reed; A2, dense stand of reed and reedmace; A3, nearly closed stands of reed and reedmace with floating and submerged vegetation (*Salvinia natans, Ceratophyllum demersum, Lemna minor, Spirodela polirhiza*); A4, open stands of reedmace and reed with floating and submerged vegetation; A5, mosaic of isolated and aggregated clones of reed and reedmace with floating and submerged vegetation (*Nymphaea candida, Nymphoides peltata, Ceratophyllum demersum*); A6, mosaic of isolated and aggregated clones of reed, reedmace, bur-reed (*Sparganium erectum*), and lotus (*Nelumbo nucifera*) with floating and submerged vegetation (*Nymphoides peltata, Salvinia natans, Nymphaea candida, Ceratophyllum demersum, Trapa natans, Lemna minor, Netellopsis obtusa*); A7, mosaic of isolated clones and fringelike stands of reedmace with a small part reed with floating and submerged vegetation (*Ceratophyllum demersum, Trapa natans, Salvinia natans, Nymphoides peltata*); A8, isolated clones of reed and reedmace reed with floating and submerged vegetation; A9, isolated clones of reed and reedmace reed with rare patches of lotus; A10, isolated large clones of reed; A11, closed monospecific fields of lotus; A12, more open fields of lotus with bur-reed and isolated clones of reed and reedmace; A13, fields of floating and submerged vegetation (*Trapa natans, Nymphaea candida, Nuphar lutea, Nymphoides peltata, Ceratophyllum demersum, Potamogeton pectinatus, Netellopsis obtusa*) with rare clones of reed, reedmace, lotus and bur-reed; A16, fields of submerged vegetation (*Ceratophyllum demersum, Potamogeton pectinatus*) with rare patches of floating vegetation and sparse clones of reed; A17, continuous fields of submerged vegetation with some floating vegetation (*Ceratophyllum demersum, Potamogeton pectinatus, Trapa natans, Nymphaea candida*), bur-reed and sparse clones of reed; A18, open water fields of submerged vegetation close to the bottom.

Figure 15.1 Map of bottom relief forms in the northeastern Caspian Sea and old coastlines, compiled from pictures taken from Soyuz 22, September 19, 1976. The letters a, b, and c mark features compared among figures. Key to bottom relief forms: 1, underwater ridges (1a, of supposed wave-formed origin; 1b, ridges now forming); 2, banks and sandbars of complex origin. Numbers 3–9 comprise a series of underwater ridges and interridge depressions formed by waves and wind-induced surges: 3, sharp crests of relict ridges and those now forming; 4, underwater crests of ridges now forming at the water surface; 5, underwater crests of ridges now forming at significant depths; 6, underwater crests of ridges in early stages of formation at significant depths; 7, low slopes of underwater ridges; 8, interridge depressions with hollows now forming; 9, interridge hollows in early stages of formation. 10, Contemporary erosional hollows formed by wind-induced surges; 11, relict channels shown in contemporary relief; 12, relict channels separated by special bottom vegetation.

Figure 17.6 Surface water temperatures deduced from AVHRR data collected July 31, 1995, indicate upwelling off the eastern coast of the central Caspian. Note that the temperature scales are different for the two views. See the text for details of temperature estimation method from AVHRR data.

Figure 18.2 Geomorphological map of the southwestern Caspian Sea coast, compiled from field observations and 1980–1996 space images. Mountains: 1, denudation–erosion mountains; 2, denudation–erosion hills; 3, arid–denudation hills; 4, mud–volcano hills; 5, deluvial–proluvial washout slopes; 6, intermontane valleys. Plains: 7, abrasion–proluvial terrace (Q2, Q3); 8, abrasion–accumulation marine (Q3); 9, alluvial–proluvial submontane (Q3); 10, proluvial–deluvial submontane (Q3); 11, alluvial–lacustrine–deluvial (Q4); 12, lacustrine–solonchak (Q4); 13, alluvial (Q3, Q4); 14, delta (Q3, Q4); 15, accumulational marine (Q4); 16, accumulational marine, modern-day; 17, accretion. Subaqueous marine plains: 18, shallow-water active wave zone and river-sediment flare distribution zone; 19, extreme coastal swell zone; 20, very deep area, below wave base. Coastal types: 21, erosional; 22, depositional. Relief forms: 23, fractures (lineaments); 24, ledges and sharp inclines; 25, mud volcanoes; 26, detrital cones; 27, erosion ridges, ravines; 28, old river beds; 29, uneven sands; 30, segments of modern-day coastal washout; 31, coastal barriers; 32, lagoons; 33, plumes of suspended material removed by rivers. [*Editor's note on terminology:* Deluvium refers to slope material, creep deposits, talus, etc., usually transported under periglacial conditions. Proluvium refers to mudflow and flash flood deposits of a bajada-type alluvial fan (*Fairbridge Encyclopedia of Geomorphology,* p. 678).]

Figure 19.2 Space photograph of the Volga delta region and the coastal zone of Kalmykia, April 14, 1994 (NASA photograph STS059-218-64). The white box shows the area featured in Figure 19.3.

Figure 19.3 Space photograph of the Lagan coastal region, September 1994 (NASA photograph STS064-101-17).

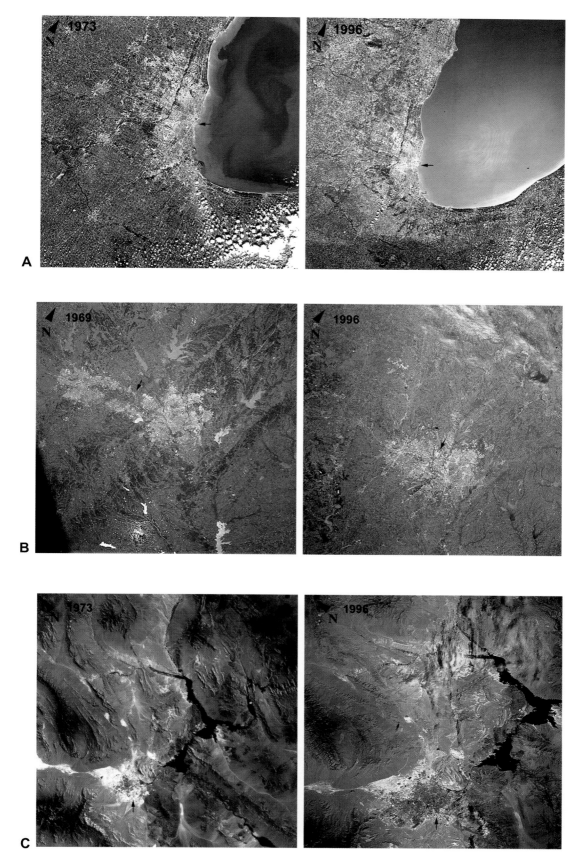

Figure 3.1

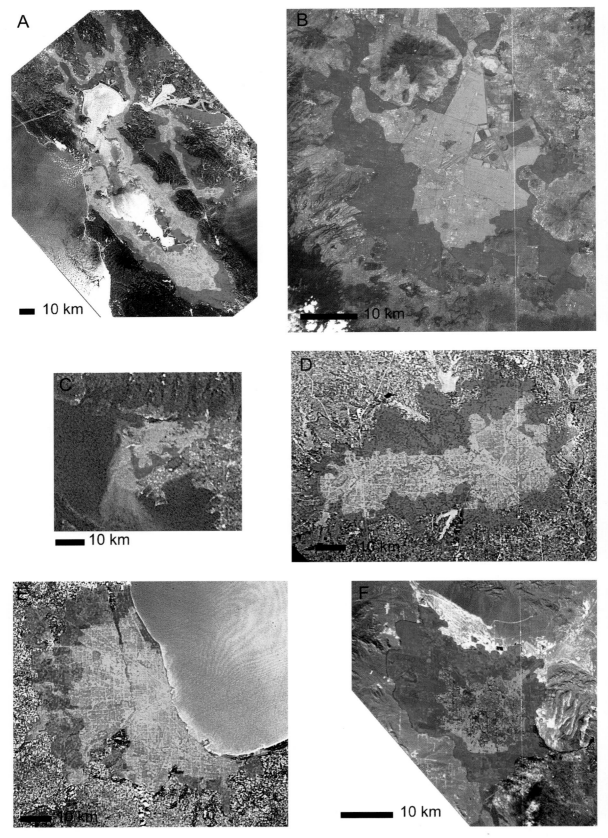

Figure 3.5

Figure 4.2

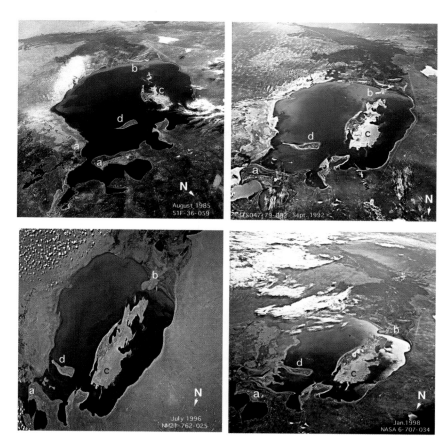

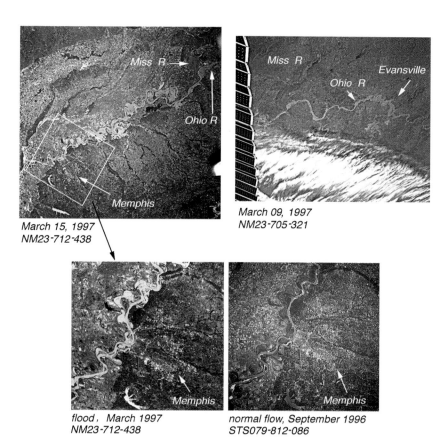

flood, March 1997
NM23-712-438

normal flow, September 1996
STS079-812-086

Figure 4.5

Figure 4.8

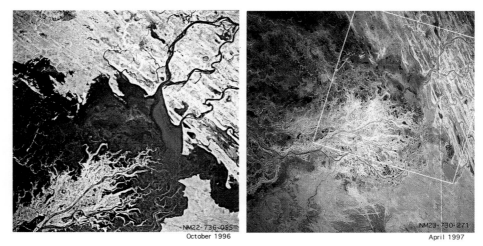

NM22-736-085
October 1996

NM23-730-271
April 1997

Figure 5.1

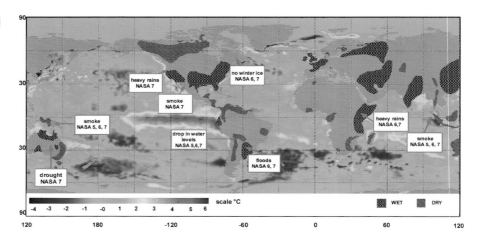

Figure 5.4

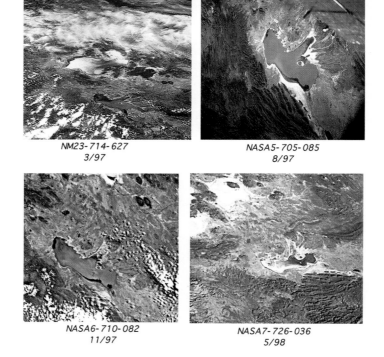

NM23-714-627
3/97

NASA5-705-085
8/97

NASA6-710-082
11/97

NASA7-726-036
5/98

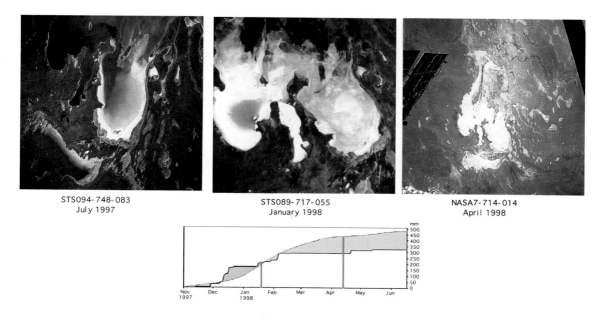

Figure 5.7

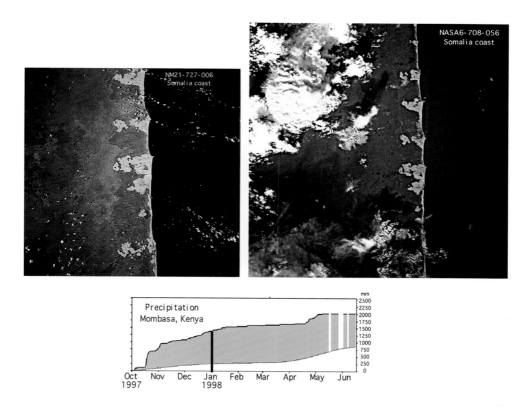

Figure 5.8

Figure 5.9 NASA7-709-029

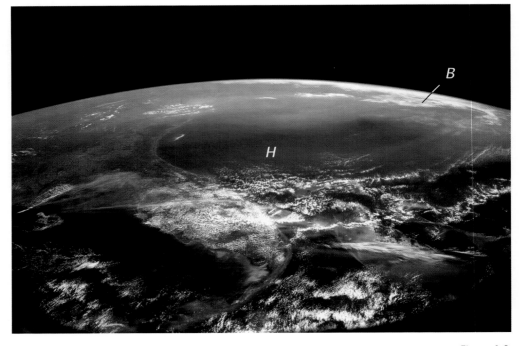

Precipitation
San Francisco, CA

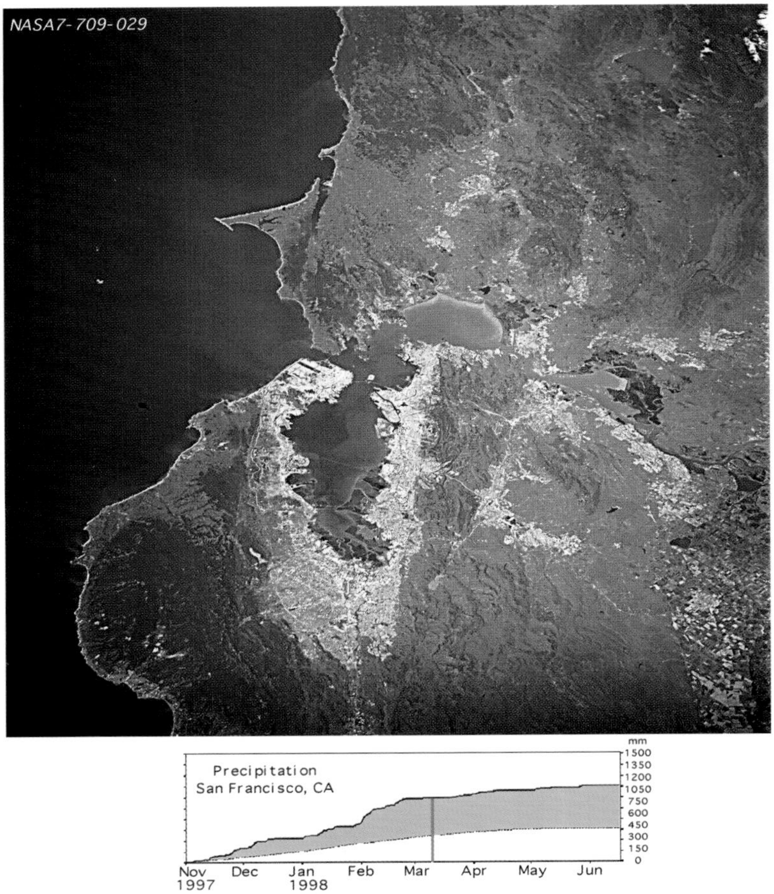

Figure 6.3

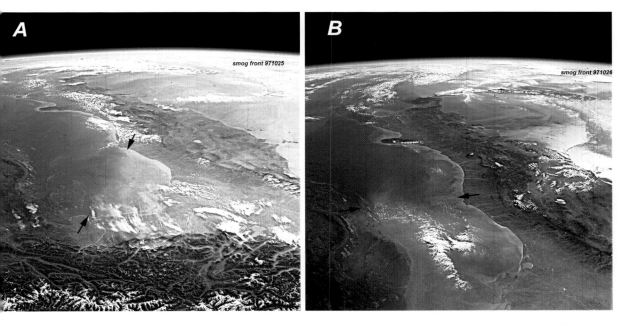

Figure 6.8

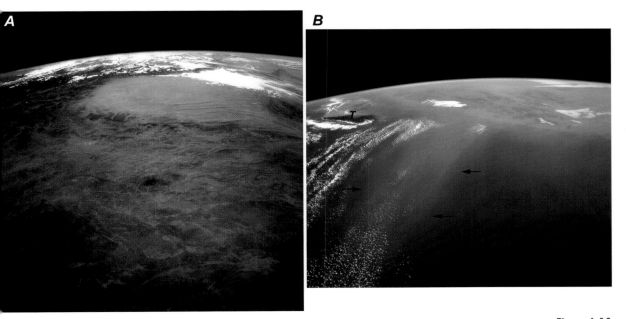

Figure 6.10

Figure 6.11

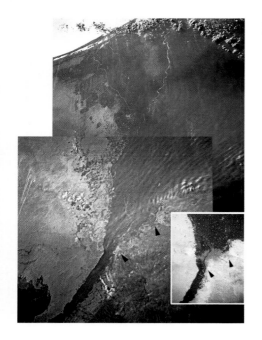

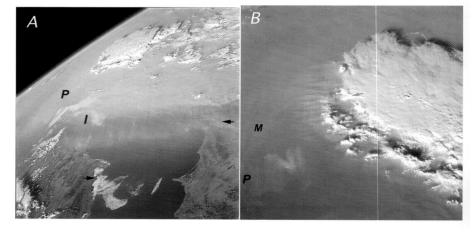

Figure 6.14

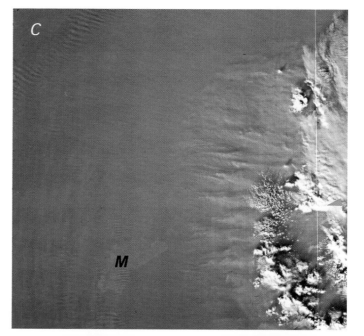

Figure 6.16

Figure 6.17

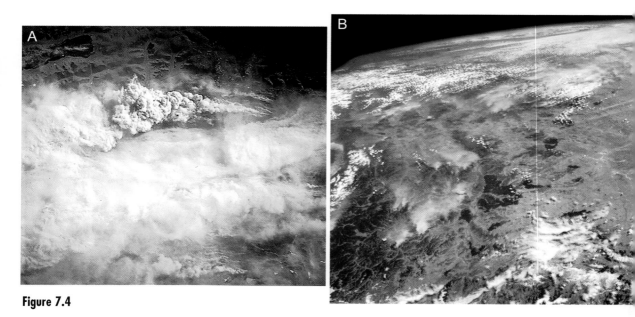

Figure 7.4

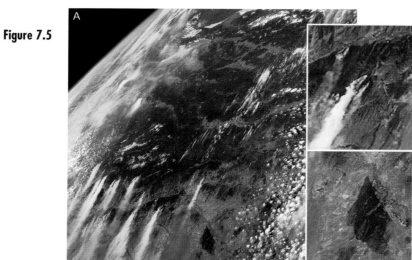

Figure 7.5

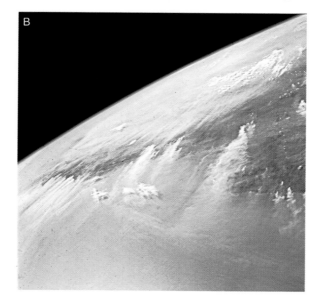

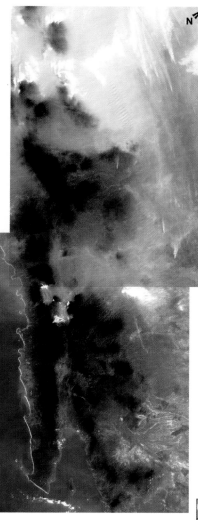

Figure 7.6

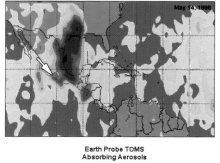

Figure 7.7

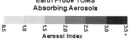
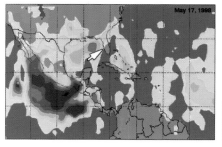
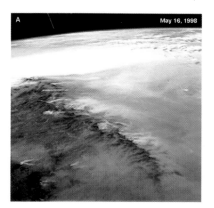
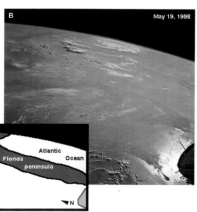

Figure 8.1

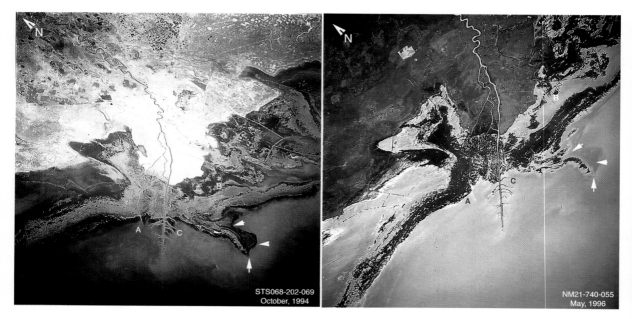

STS068-202-069
October, 1994

NM21-740-055
May, 1996

Figure 10.3

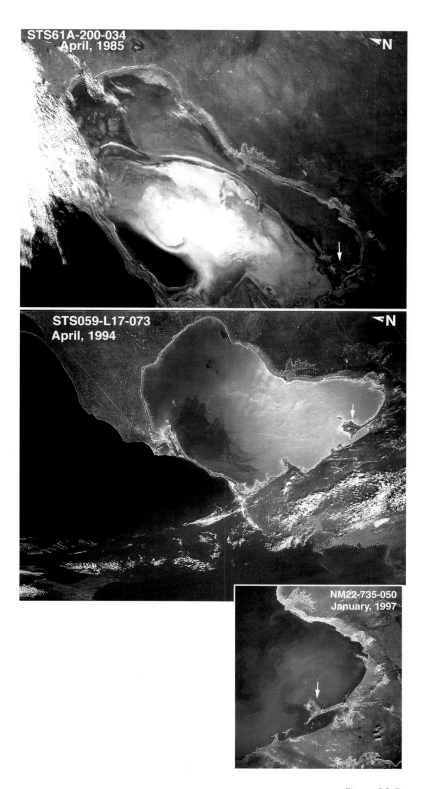

Figure 10.5

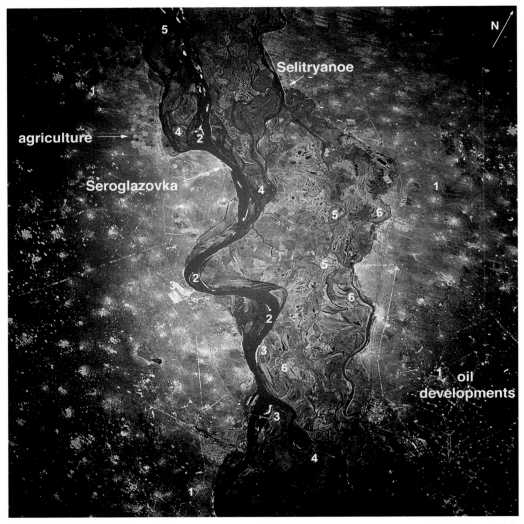

Figure 11.6

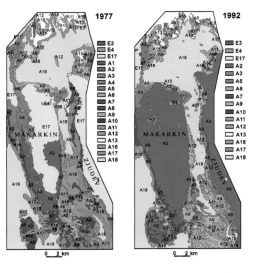

Figure 13.4

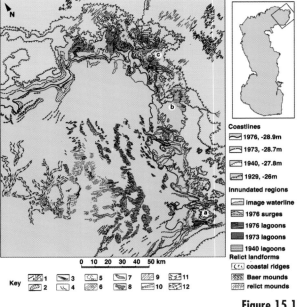

Figure 15.1

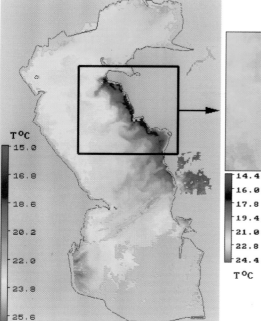

Figure 17.6

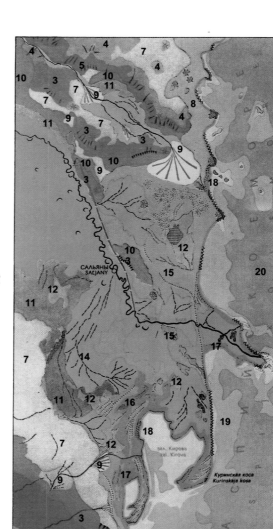

Figure 18.2

Figure 19.2

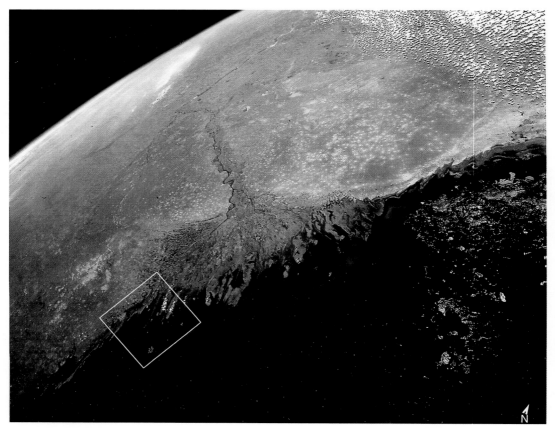

Lagan

STS064-101-017

Figure 19.3